TEN TRIPS

TEN TRIPS

The New Reality of Psychedelics

ANDY MITCHELL

HARPER WAVE

An Imprint of HarperCollins*Publishers*

HarperCollins books may be purchased for educational, business, or sales promotional use. For information, please email the Special Markets Department at SPsales@ harpercollins.com.

Originally published in the United Kingdom in 2023 by The Bodley Head.

Epigraph from Jorge Luis Borges and Osvaldo Ferrari, *Conversations, Volume 1*, tr. Jason Wilson (Seagull Books, 2014).

Lyrics quoted on page 153 are from "Strawberry Fields Forever," written by John Lennon and Paul McCartney (1967); "A Whiter Shade of Pale," written by Keith Reid, Gary Brooker, and Matthew Fisher (1967); "What a Wonderful World," written by Bob Theile and George David Weiss (1967).

FIRST U.S. EDITION

Library of Congress Cataloging-in-Publication Data has been applied for.

ISBN 978-0-06-322038-6

23 24 25 26 27 LBC 5 4 3 2 1

For my daughters Lois and Dulcie

Last week . . . two people asked me the same question – what's
the use of poetry? And I answered them with: What's the use of
death? What's the use of the taste of coffee? What the use of me?
What's the use of us?

Jorge Luis Borges to Osvaldo Ferrari,
Conversations, Volume 1

Contents

Introduction

In recent years, after a lengthy hiatus, conventional medicine has once again taken a psychedelic turn. Research began in the 1990s, rekindled by a few isolated programmes in Europe and the US. In the US the first scientific experiments with psychedelics in over twenty years were conducted by University of New Mexico professor of psychiatry Rick Strassman. Between 1990 and 1995, his government-backed programme investigated the effects of an extremely powerful psychedelic that occurs naturally in a large number of plants and mammals, known as DMT, 'the God molecule' – for its association with spiritual experiences – or, more affectionately, as 'the Businessman's Trip', on account of its disguisable brevity.

By the 2010s, a growing body of researchers were focusing their attention on mental health, a field many regarded as in desperate need of advances. Over the course of the decade, a series of studies conducted by various research teams indicated the successful application of psychedelic drugs in the treatment of anxiety and depression in cancer patients, as well as alcohol and tobacco addiction and otherwise treatment-resistant depression. These ground-breaking studies went hand in hand with a relaxation of regulatory prohibitions and increasing public curiosity, aroused by the largely positive reporting of the media. The year 2018 proved to be something of a psychedelic *annus mirabilis* with the publication of Michael Pollan's careful, compelling bestseller *How to Change Your Mind: The New Science of Psychedelics*,[1] and the US Food and Drug Administration designated psilocybin (magic mushrooms) as a 'breakthrough therapy' in the treatment of depression.

Since then, numerous well-funded research teams across several continents have conducted multiple studies and clinical trials with various psychedelic drugs. Between them they have delivered remarkable insights into the effects of psychedelics on the brain and the potential treatment of a range of mental health disorders. Medical research has rippled out well beyond the field of psychiatry, with a growing number of studies published on psychedelics as a possible treatment for conditions ranging from Alzheimer's disease to aneurysms, chronic pain, eye disease, inflammation and immunological problems. This kind of research now routinely makes headlines in mainstream media: 'Will the magic of psychedelics transform psychiatry?' asked the *Guardian* in November 2021. A year later, 'Psychedelic drugs may launch a new era in psychiatric treatment, brain scientists say,' according to NPR, and the implication was, with increasing confidence, that they would.

And with the buzz come more investment, more innovation: 'Wall Street Backs New Class of Psychedelic Drugs,' reported the *Wall Street Journal* in February 2023. Venture capital funds have created multiple international start-ups, such as COMPASS Pathways and Atai Life Sciences, who compete to develop and patent variants of psychedelic drugs and therapeutic protocols to sell to healthcare providers for widespread medical use. In the US alone the projected value of the magic mushroom market in 2028 is $6.4 billion, which puts it on a par with that for baby food, and nearly ten times that of M&Ms.

All of which is to say that in a few short years psychedelics have gone from something you did at a festival to a place in the cultural mainstream. Being an expert in them had meant being a psychonaut: something of an outsider, of less than exalted reputation. Nowadays it means being a research scientist, a therapist, an investor. What was for decades marginal has once again become of interest to the ordinary and extraordinary alike: in March 2023, Gabor Maté, the well-known physician and author specialising in trauma and addiction, interviewed and diagnosed Prince Harry on television, following the Royal's astonishing, transformative experiences with a psychedelic tea.

Like some of my younger, more progressive colleagues, I kept half an eye on these developments. It's territory that's familiar to me professionally as a clinical neuropsychologist, treating people with brain trauma, neurological conditions and, before that, mental illness and addiction. At the time there was little in the way of personal interest: in my teens and twenties I'd taken more than my fair share of drugs, including the occasional trip (it was a long time ago, the experiences were 'messy', more torturous than recreational, and I had no desire to revisit them). But in middle age the things I'd replaced drugs with – mindfulness practices, an interest in personality change (my own included) and self-transcendent states – were the very things psychedelic medicines were now being associated with. And yet, impressive as some of the studies undoubtedly were, some of the claims being made for psychedelics made me increasingly uneasy. For a start, the science did not always seem especially robust. The best scientists were open about the fact that their findings were qualified or partial or preliminary. Less scrupulous ones brushed over the fact that their studies were based on small cohort sizes and had yet to be replicated. Often I had the feeling that their attention was in the wrong place.

The resurgent interest in psychedelics has concentrated primarily on the brain, and more specifically on those areas of the brain currently in vogue (the brain has its fashions too), and very little on the actual experiences of the people taking the drugs. Or else those experiences are binned in broad, coarsely constructed categories like 'mystical experience' that reflect the mindsets – the biases and/ or preferences – of the researchers themselves. I read reports about the sudden, dramatic and lasting treatment of mental illness. But my experience as a clinician has been that mental illness doesn't tend to respond like that.

Clearly there was and is something incredibly interesting and valuable to be understood scientifically, but also culturally. Like many, I suspect, I had friends, or friends of friends, who had tried psychedelics, out of curiosity, but also as a way of addressing feelings of misery, anxiety or stuckness. But I also couldn't help noticing how interesting and valuable psychedelics were to investors and had questions about the kinds of investment being made, and in whose interests; fashion and science are often uneasy bedfellows. From my perspective – the outside, that is – it seemed we had an unhealthy cocktail of 'rigorous expertise' (which was perhaps not quite as rigorous or expert as it might seem) and hype for commercial gain.

This book sets out to pose some big questions. Who gets to speak authoritatively about psychedelics? Who actually *knows* about them? We in the West, where they remain illegal almost everywhere, or the people who have a living, ongoing cultural relationship with them stretching back thousands of years? And if it is us, is it the scientists or the therapists or the New Age seekers or the investors or the psychonauts? How separable are these factions? And, at the risk of asking an unanswerable question, what does it even mean to know something? Is a scientific paper knowledge? Is cultural wisdom knowledge? Is experience itself knowledge, or does it depend on a model of health economics? One thing did seem clear to me: between the scientific literature, the commercial ventures and the experiences themselves (in my case only vaguely remembered or insufficiently repressed from twenty years back) there were significant disjunctions. Could these ever be bridged – could those worlds come into contact with and understand one another, and speak the same language?

It turns out that psychedelics are the test case par excellence for thinking about the relationship between science, culture and commerce, and the wider questions I've posed, precisely because the experience of taking them is so various, so personal and so often simply inexplicable. As will become apparent, they are slippery, tricksterish, risky in different ways. They may be 'mind-' or 'soul-manifesters' (in the ancient Greek, *psyche* can be roughly translated as 'mind' or 'soul', and *deloun* 'to make clear or manifest'), but they are also illusion-mongers that defy generalisation. One of the things that struck me most was how the powerful claims being made for psychedelic drugs seemed to come with a simultaneous attempt to sanitise them, to categorise them as safe, normal, acceptable. And yet the people I would come to know who'd taken them often didn't think of them like that at all.

But I'm getting ahead of myself. Back in late 2021 (I'm writing this in early 2023) that mixture of professional interest and uneasiness bordering on scepticism was as far as it went. I had no intention of taking psychedelics myself, or indeed any other drug. Until, that is, I met Aurora Valdez, a friend of my cousin – he thought we'd have an 'interesting conversation' – who within minutes of our meeting was telling me she was an *ayahuascara*.

This was in California – Big Sur, in fact, garden of hippy culture in the 1960s and 1970s, where such confessions are not uncommon, I assumed. 'You've heard of ayahuasca?' There was an intensity under her casualness. She had a huge pendant round her neck, whose Aztec design was rendered in a thousand tiny neon beads.

'Sure,' I replied. The organic Amazonian psychedelic, famous for the intensity of its experience.

'Well, an ayahuascara is the person who leads the ceremony. That's me. Serving the tea . . . You're looking dubious.'

I was. I'd been clean and sober for a couple of decades now – no drugs, no booze, just kombucha. It suited me that way.

We were having lunch in an old ranch-style café whose clientele was an unlikely mix of Asian tourists, high-functioning hippies, precocious, techy second-homeowners (Silicon Valley had spread this far south) and antique pioneer types, their faces creased and weathered like horse saddles.

'It's not *drugs*, you know,' said Aurora. 'It's a sacrament. *Medicine* is what it is. An ancient indigenous tradition with a lineage of *icaros* – beautiful songs inspired by the plants themselves. *Healing* is what it is.' As a clinical neuropsychologist I ought to know that psychedelics have astonishing therapeutic benefits that have been clinically documented.

She asked if I'd heard about MDMA's dampening effects on the amygdala.

I had.

Or what mushrooms can do to the default mode network of the obsessive and the depressed . . . Or how micro-doses work as cognitive enhancers . . . The neuroscientific vocabulary *flowed*. She was speaking my language – everyone was these days – from behind sharp green eyes.

As she was speaking I was struck by the thought that her pupils looked dilated. Was this meeting *enhanced*? I glanced round the restaurant: it felt as though everyone was more engaged, more animated, less self-conscious than you'd expect. Maybe Aurora had chosen this restaurant because it was an ayahuascoid hangout. Or, worse, maybe this was just normal now, and I was the last naïf in the state: a douchey forty-nine-year-old psychedelic virgin.

'Have you read *How to Change Your Mind*?' Aurora asked. Michael Pollan's ground-breaking, culture-creating book had recently been made into a four-part Netflix series whose message was clear: psychedelics have the capacity to revolutionise the way we understand and treat mental health. Of course I'd read it; it was great. It just hadn't quite changed my mind.

But then, *my* mind was in questionable shape. What Aurora didn't know – what I hadn't told her – was my recent 'clinical' history. Over eighteen long and devastating months my friend Danny, who had worked with me in the traumatic brain injury unit at a central London hospital, had his life first dismantled and then extinguished by amyotrophic lateral sclerosis, a horrific variant of motor neurone disease. At the same time my eldest daughter had become ill with another terrible condition that only seemed to get worse the more treatment she received. Then, out of the blue, my father died. I had taken compassionate leave and come out to the West Coast of America for ten days to see old friends and pull myself together.

In the world of psychedelics, one's 'set' – the beliefs, values and preferences of the consumer – has a crucial, even defining influence on how the drugs affect you. It would be fair to say that mine was a combination of fear and scepticism about the drugs added to a more general grief and fragility. One's 'setting' – the environment in which the drugs are consumed – is also crucial to the psychedelic experience, and in this case it was Big Sur, California, where natural law is a little looser, a little less fastened down.

As Aurora and I walked along the beach after lunch, I rehearsed my general objections with her.

'Your prejudices.'

'My reservations.'

'Your preconceptions.'

'All right then, my preconceptions.'

I told her how these days depression kills more people than heart disease and cancer. At some point in life one in four adults will meet the criteria for a mental health diagnosis. For the first time in modern history suicide is lowering general life expectancy; the diagnosis of post-traumatic stress disorder is doubling every few years. And then we've had the pandemic. Meanwhile, our populations are ageing, and old age itself is a negative indicator of mental well-being, even while it is the young who seem hardest hit. For decades, there has been no pharmacological innovation in psychiatry, not since the introduction of the most widely prescribed class of antidepressant SSRIs (selective serotonin reuptake inhibitors). These have been valuable up to a point, even if their efficacy in controlled trials is hard to distinguish from placebo. And now, suddenly, the botanical superhero makes an appearance: *Antidote!* The timing was uncanny. It was perfect.

'And that's a bad thing, is it?' said Aurora. 'Perfect timing?'

Of course not, I explained, but it might also be a symptom of wish-fulfilment: engulfed by multiple existential threats, psychiatry all but washed up, and suddenly we rediscover this seam of indigenous atavistic knowledge to save our skins in the nick of time. The Sixties nemesis of mental health, repackaged as the nemesis of contemporary mental illness.

Aurora demurred expertly: according to this logic, she said, I would have mistrusted penicillin because they managed to mass-produce it just in time to help win the Second World War. But I still couldn't help thinking of the hype machine. How often our lives as consumers are shaped by the illusion of choice: Coke or Pepsi, Insta or TikTok, petrol or hybrid, Exit or Dignitas, cow's milk or oat. Or, here in central California, almond or coconut or flax or quinoa or hemp. And now it's SSRIs vs psilocybin. One makes you feel dead inside, as many of my patients have complained, the other may give you a whole life in a single night, for better or worse. And with psychedelics it's more than just popping a pill. You can take SSRIs sitting on the toilet in your boxers without a thought in your head. But because with psychedelics the 'set' and 'setting' matter so much, making the treatment simultaneously pharmacological and psychotherapeutic, elaborate and bespoke therapeutic protocols have to be developed – a whole 'extra-pharmacological' theatre, furnishing yet more opportunities, in the form of patents and IP (intellectual property), to commercialise them. 'If they didn't exist,' I finished up, 'you'd have to invent them.' I was on a roll.

'But they do exist! You just haven't tried them!'

The ocean bruised as the sun dipped behind a cloud momentarily. I was nailed.

'Never mind all the experimental evidence and the capitalist-conspiracy think,' Aurora continued. 'I have first-hand knowledge of how ground-breaking these things are. They change people's lives.'

Aurora Valdez: even the name – a hybrid of New Age and Latin American – sounded psychedelic. Two hundred years ago her ancestors had walked to America from south of the equator. Meanwhile, the British – me, I mean – wear our brains on our sleeves rather than our hearts. I wasn't about to tell a stranger all that was on my plate, but I did want to feel differently. In truth, I longed for it.

I was still looking out over the ocean. The Pacific, the Amazon – is it possible to conceive of psychedelics without such infinities?

What I might have said to Aurora – what I was too preoccupied to say – was that another way of understanding the hype around psychedelics is to see them as technologies of humanism – the belief in our species' unique ability to improve ourselves and the world – rather than of dispassionate scientific method; as simply the latest players in the long drama we like to call progress. From this perspective hype is really *hope*, part of what the philosopher John Gray has called the prevailing Western secular world view – improvement, civilisation, reason overcoming adversity – which is in reality 'a pastiche of current scientific orthodoxy and pious hopes'. It's no small irony that psychedelics – once totems of the unreasonable, of irrationalism, of social degeneracy – have become the drug of choice for human reason and well-being. The dark mirror of the American dream reconfigured as that same dream's enabler. You could have your vegan space cake and eat it. In fact, you *should*. And of course, for all their mind-manifesting promise, psychedelics may also be what drugs always have been: merely the instruments we use to escape an unbearable reality.

I could have said all of that, but instead I said, 'Tell me about the dosages. What's the level of risk?'

Aurora explained that the risks were statistically small within a properly administered setting, but that there *were still* risks that could not be completely contained or offset. And she said it so calmly, so reasonably, so soberly. At least one of us knew what they were talking about. For the rest of the afternoon I listened as she laid out her stall, gently and eloquently pushing her wares on me.

The following day, I called my therapist, Nigel, back in London. Only 15 per cent of Americans have a positive opinion about using psychedelics, I told him; 34 per cent have a negative opinion, and about half are neutral. Despite my framing, he thought that sounded like a large number of undecided. Worth exploring, at least. But then Nigel was a child of the Sixties; he would say that. I called Lenny, my AA sponsor. He told me that in the Fifties Bill Wilson, one of AA's founders, had had two years of LSD therapy which by all accounts had totally transformed him. Lenny had no material objection; I'd been sober a long time, long enough to make good decisions. Taken the right way psychedelics were a form of spiritual enquiry, and AA was a spiritual programme. But, he reminded me, I had had a 'Full English Breakfast of grief' recently; whatever I decided, I should go carefully.

A few evenings later, I'm standing on a cliff top outside a ceremonial yurt in Julia Pfeiffer Park shouting into my phone, '*THIS!*'

'Hello?'

'*THIS!*'

Half-a-dozen low-flying angels have just whooshed past my head. What I'm trying to say to Nigel is something like, '*I'm having a zero-gravity hyper-gasm'*, or, *'I've just discovered Black Ball for the human heart'*. It's less than a week since my lunch with Aurora, where I had made my position very clear. Now my position is: *very, very high*.

'Is that Andy?'

'Just, *THIS* . . .'

It was nearly four hours since Aurora had administered the tea, a khaki-green shot that tasted like someone else's bile. Twelve of us had been spaced out around the perimeter of the yurt like hours on a watch. One hour later, we were all sliding down a melting time-piece like the numbers in Dali's *The Persistence of Memory*, while our ayahuascara sang of ancient Peruvian doctors healing the sick with sacred plants. And then I was sick. One moment I was in a terrible crevasse of regret, wondering what drunken self-delusion had per-suaded me to actually drink this stuff; the next: *'This . . . This is the future, and it's three thousand years old.'* I wanted Nigel to understand the time-collapsing, identity-destroying oneness of all things, approximately speaking. But all that came out was, *'THIS!!!'*, which meant gradually coming to terms with the fact that the trip experience had a built-in condition: it defied description. Nothing could be brought back from the other side, other than the (mis-placed) conviction that everything was wondrously self-evident. *'This . . . This.'*

The ocean continued to breathe on my behalf, the night-sky fold-ing itself like a letter.

'Andy, are you OK . . . ? Do you have a question?' Nigel's voice crackled over the sound of people vomiting inside the tent.

'. . . Hey, dude, if *"this"* is a question,' I said, 'then *"this"* is the answer.' Words were returning as the clarity of my vision was beginning to smudge.

What a night! I spent most of it in the company of my dead father, my late colleague Danny and my eldest daughter, their minds and bodies beautifully restored. From one moment to the next it was heartbreaking and terrifying and funny. It felt more real than waking life. At one point in the journey's deepest moment, I opened my eyes and Aurora was standing over me singing in Quechua, her body, dressed in a brilliant white shroud, stretching upwards like astral light into the far reaches of space, her head swallowed by limitless dark. Slowly, she raised her arms aloft, and the earth beneath me trembled and groaned. This is what real healing feels like, I thought.

For all my bluster I was, at the level of experience, utterly ignorant. More to the point, I had no idea *how* ignorant. My whole life felt cut in two either side of the experience. Its wonder was matched only by its terror, its loopiness by its precision. It exceeded anything I had imagined by orders of magnitude. And that was the point: not only was it 'inconceivable', but it also made me feel as though something else was doing the conceiving – which was both unsettling and strangely relieving. At the same time it felt unmistakably *mine*, shaped to fit my 'set'. It transformed my relationship to my dead father, allowing me to say to him what had remained unsaid and heal a rift that had been inaccessible in real life. It also allowed me, after two pain-filled years, to see something new about my daughter's illness, that it was a kind of koan uttered by 'the Universe' (it was California) according to which the more I tried to help her, the worse her condition became.

And so the seed was planted that grew eventually into this book.

Ten Trips is a journey through ten different 'settings' – from neuro-imaging labs through seven-star psychedelic retreats to the mountains and forests of Colombia. In each of the ten locations I took a different psychedelic drug (with one exception), covering all the classics and some less well-known, and through them brought my personal and professional experience to bear on the major narratives of the Psychedelic Renaissance.

The ten chapters in this book correspond broadly to these ten trips. I start in the West, in medical and therapeutic settings, and slowly make my way south, towards more historic, indigenous settings. I'm a little loose with the traditional classifications of psychedelics, including relatively recent laboratory constructions (ketamine, LSD, MDMA) as well as ancient plant brews (*yage*, *wachuma, iboga*). The book begins, at least, as an investigation and a survey, an argument and a treatise on the scientific/therapeutic understandings of psychedelics as medicines (sometimes it will get a little technical, but stick with me), in part to scrutinise and perhaps justify some of my scepticism about the more extravagant claims being made for psychedelics. But more importantly it's an attempt to widen our field of vision, to see what's interesting and valuable about psychedelics that the disciplines of science and therapy seem to neglect.

It's also a faithful reflection of what I discovered as I went, and what I discovered does not fit comfortably into the bounds of a sober report. In fact, I had more 'journeys' in fewer days than any responsible doctor would recommend. For some doctors, and for many people, any drug-taking – whichever drug, whatever the circumstance – is too much. This book is likely not for you. Either way, be prepared for contradiction, inconsistency and some recklessness, which inhabit the different perspectives and uses of psychedelics themselves. But they also reflect the fault line(s) in my own character. Sometimes I meet my experiences as something like an 'inquisitive clinician' – think Oliver Sacks on a bad day – at others, in a less professional, less responsible (one might say less 'legal' if the whole shebang hadn't been almost universally criminalised) way – think more Hunter S. Thompson on a good day. Either way, please don't worry unnecessarily for my patients' safety. I can no longer be struck off: for the last few years my clinical work has only been piecemeal and voluntary.

Many of the chapters incorporate other trips and experiences, too, following more thematic threads. Overall, they trace a trajectory from the medical and scientific to the religious and spiritual, to the cosmological and existential, to the anarchic and playful and . . . well, wait and see. Along the way, I bump my head again and again on the question of how we perceive reality, and even what we mean by those two concepts, perception and reality. Suffice to say that the book begins thinking about where we might be going wrong, and ends by thinking about how we might make good on our mistakes – how we might think better about these drugs, and maybe *with them* too on occasion.

Because these days the psychedelic field resembles the site of a rave – vast, over-trodden and lacking signage – I hope this book can offer the reader a handrail, or at least a rough map, albeit at a scale which may be misleading, with crucial markings disappearing into the creases. It is in some ways a continuation of Michael Pollan's *How to Change Your Mind*, an updating of the conversation five years on, as well as a riposte to some of its orthodoxies, including an extra dose of wildness. A guiding metaphor throughout *Ten Trips* has been that of psychedelics as a multi-faceted gem: depending on the viewer's vantage point, a different facet of the gem is reflected. Any single perspective is by definition limited, and therefore unable to apprehend the gem in its entirety. To date the Psychedelic Renaissance has largely been the preserve of medical science and corporate investment, which has led to factionalism and the relegation of non-medical views. But an authentic renaissance suggests a wholesale flourishing that includes the arts and humanities, law, politics and economics alongside the sciences. The aesthetics of the gem or its particular cultural meaning are just as important as its geometry.

Perspective implies a subject looking at an object. But with psychedelics the metaphor of looking breaks down (as all of our metaphors tend to). The distance doesn't hold. Albert Hofmann, the chemist who invented LSD, put his finger on this when he described reality as 'the product of a transmitter, the material, exterior world, and a receiver, our consciousness, the inner, spiritual centre of a human individual'. Each of us has what Hofmann called 'antennae', except that these antennae help to construct reality as much as they receive it. We therefore need to know something about the nature of our own antennae, and this has become another metaphor I've used to navigate the psychedelic space. Such self-consciousness is common practice in the arts and social sciences, just as so-called 'observer effects' are fundamental to quantum mechanics. But in medical science they are rarely considered. This is arguably less important for most bio-medical phenomena, where there are 'hard' signs that can be reliably validated. But psychiatry is different. And psychedelics are different, too. Their sensitivity to set and setting, and their association with heightened states of suggestibility (one aspect of their 'plasticity', to use a neuroscientific term), necessitate an uncommon degree of self-awareness when it comes to thinking about or interpreting them.

As I write, there are academic hubs throughout the US, Europe, Australasia and South America, each with their own particular areas of interest when it comes to the medico-therapeutic application of psychedelics. They elaborate different neurocognitive networks, explore the parameters of set and setting, attempt to disentangle the methodological problems of the placebo effect and 'blinding' controls – and all of it, of course, performed by scientists on co-operative patients or healthy volunteers and, for certain more invasive procedures, on animals. That's the third-person perspective of science: disengaged, objective (or attempting to be), the way of seeing we have come to know and trust since it emerged in Western Europe some 400 years ago.

But there's another, different tradition which survived well into the twentieth century – Albert Hofmann's tripping bike ride being a celebrated example – in which it was standard practice for doctors and scientists to routinely experiment with drugs on themselves: the so-called first-person perspective, in which subjective experience is rendered central. While most scientists took care to minimise risks, notes the drug historian Mike Jay, 'many were powerfully drawn to the thrill of exceeding the limits of the known, often with dramatic consequences. Reckless drug experiments and descents into madness were at the core of many of the era's most popular fictions.'[2] And yet these were not just entertainments or exercises in gratuitous risk-taking. A deeper risk, thought such scientists, was in *not* taking them, because there was something so profound at stake.

The authority of the third-person perspective depends on dispassionate objectivity, which in turn relies on isolating and quantifying variables, controlling parameters and arriving at rules that can be generally applied and argued over. In the process the particularity or quality of experience is lost, jettisoned in pursuit of a 'higher' cause. Increasingly, however, another kind of authority is making a comeback: one with the greater interest in identity that is characteristic of contemporary politics and culture. Personal experience has become once again a key qualification for being trusted, or at least properly listened to. This is, of course, an older kind of authority – perhaps the oldest – and one that cannot be generalised or argued with. Scientifically, it may look like an imbroglio, even a dead end. And yet one of the key insights I wish to share about psychedelics is that their particularity – to the moment, to the place, to the person – is everything. They resist being generalised. That's kind of the point. In fact, they resist *being* most things, while *looking like* different things to many different groups of people. And science needs to find a way of recognising this. Which is to say that to understand the conundrum fully, from both sides, it turns out that taking the drugs myself and writing about them in the first person was the best and only way to proceed.

Robin Carhart-Harris, professor of the Neuroscape Centre at University of California San Francisco, recently suggested that psychedelic hype is akin to Joseph Campbell's myth of the 'hero's journey', by which I presume he means that the whole culture has yet to undergo various stages of development before it arrives at full maturation. As of today, our hero is developmentally young, full of potential, but tending to evangelical optimism and exaggeration, and a little resistant to nuance or counter-narrative. Real growth depends on a real journey – hubris's encounter with the Fates, disappointments as well as triumphs – which takes the hero far beyond the narrow orthodoxies of the West in search of a fuller appreciation of the unbelievable beauty and strangeness of these drugs, to arrive at a more worldly scepticism, a deeper wonder, that one might associate with maturity. This, I hope, is the journey of this book.

PART ONE

I

Psychedelically Naïve

For my first trip I would receive a 'super-dose' of ketamine intravenously while having my brain scanned in a 3-Tesla fMRI machine. Unless I was lucky, in which case it would be a high dose of dimethyltryptamine (DMT), lying between the large rings of a PET (positron emission tomography) scanner. This was the psychonautical equivalent of a three-star Anthology meal at the Fat Duck, or a performance of Beethoven's Seventh by the Berliner Philarmoniker, if the meal were eaten standing up in an aeroplane toilet, the concert heard over the meleé of Black Friday on Oxford Street.

Ketamine: *special K.*

'It's a *mong* for end-stage burners,' said Palmer, techno DJ and medicinal gourmand. 'I could play Abba and no one would clock it.'

'It's a horse tranquilliser,' said my drug-naïve mum. 'Those Thai soccer boys took too much and got stuck in a cave for days.' (She'd only *heard* about the Netflix documentary.) 'I think the rescue team gave them it to stop them panicking, Mum.'

'It's a Swiss Army knife,' said an anaesthetist colleague, 'used offlabel as an anti-inflammatory, for pain relief, for neuroprotection, as well as an anaesthetic in surgery and critical care.'

I'd also read about anti-ageing hacks in California 'playing' with mega-doses that led to a complete collapse in space–time coherence, a few minutes stretching out to a 'felt' century. I would be catheterised for the best part of an hour. It might be that the first human to live 5,000 years (me) had recently turned fifty and already had flattening arches and an arthritic hip.

These perspectives layered my ketamine 'set' – my 'priors', to use the jargon of the neuroscientist, meaning the beliefs one holds before any experience has taken place, and the tendency for those beliefs to shape the experience itself.

The study was being conducted by Imperial College's Centre for Psychedelic Research, an international leader, and one of the first groups to be established with seed money from the philanthropist/investor/podcaster/author Tim Ferriss, of *The Four-Hour Work Week*, *The Four-Hour Body* and *The Four-Hour Chef*, fantasies of compression enabled, I imagined, by the 4,000 years he'd spent exploring these things in psychedelic space-time. I had just received the patient information sheet (PIS), following a two-hour interview with one of the research assistants that had covered my psychological history, my educational history, my relationship history, my drug and alcohol history. And this was just the *pre*-screen. In the days to come I would have a formal clinical interview lasting several hours with a consultant psychiatrist. It made sense to be careful about who one loaded in the barrel of an MRI scanner and shot into unimagined realms. In most psychedelic trials there were general criteria for 'healthy': no history of suicidality, psychoses, bipolar, personality disorders or long-term drug addiction. And for this particular trial there were extra criteria: that I was both *ketamine-naïve* and hadn't been near psychedelics in three months, the latter being why I had elected to make ketamine the first of my ten trips.

The PIS was long, detailing every stage of the investigation in language that was supposed to be accessible to the layman to ensure the study's safe passage through the ethics committee. It didn't begin promisingly: '*Detecting synaptogenesis induced by Ketamine/Dimethyltryptamine and motor learning using the tracer [11C] UCB-J in an integrated PET-fMRI paradigm.*'

'The brain's ability to reshape and make new connections during adulthood,' it continued, 'is essential to our ability to learn new skills and form and access memories. This process, broadly described as neuroplasticity, can be disrupted by many different factors and is increasingly believed to be centrally involved in a number of mental health disorders and cognitive impairments.'

This related to the 'synaptogenesis' part of the title. New

experiences may be registered in the brain by the generation of dendritic spines which sprout, tree-like, on one end – the dendron – of the neuron. The language of neuroplasticity is infused with the metaphor of the tree: the sprouting is called 'arborisation' after the Latin; under high magnification the dendrites look like foliage. New Age psychedelic therapists like the recently disgraced Françoise Bourzat take the metaphor a stage further, seeing in the images of tripping brains anatomical symbols of the plants or mycelial (fungal) networks that inspire them. To the more circumspect this might be no different from getting high and seeing the profile of Donald Trump's quiff in wispy clouds. Most neuroscientists would think of both as examples of pareidolia, the tendency to perceive a specific, often meaningful image in a random or ambiguous sensory pattern. Bourzat's gloss on ancient human–plant synergy works less well for ketamine, which has no botanical basis but was developed in 1962 in a Detroit lab owned by a subsidiary of the pharmaceutical giant Pfizer; a little more difficult to romanticise than Mazatec mushrooms and Amazonian vines.

Neuroplasticity has for decades been one of the most popularised areas in neuroscience, long pre-dating the current interest in psychedelics. 'Re-wiring' has become part of the vernacular of life coaches, football managers, schoolteachers and mobile apps. (One might describe its recent ubiquity as the 'Joe Dispenza Effect', after the chiropractor-turned-neuroplastic guru.) Plasticity is seen as a 'hack' allowing us to acquire new skills and knowledge, change old patterns or 'priors', *re-think* ourselves. (Today Joe's tweet reads: 'To create a new personal reality – a new life – we must create a new personality. We must become someone else.'[1]) This mode of understanding makes it inevitable that plasticity and psychedelics are saddled together: two 'new' instruments of improvement, passwords for the near-limitless possibilities of self-transformation, couched in the sexy-sounding language of neuroscience.

Keen to learn something of what might be about to happen to my brain, I run a Google search for 'evidence of imaging of synaptogenesis'. It yields little. A few lead-in adverts (dictated by the engine's *plastic* algorithm) for how to train your brain to give up sugar or 'speak proper grammar', then a low-res black and white clip lasting a few seconds that resembles the beginnings of cinema: dendrites like tiny forks of lightning across a night sky, appearing, then disappearing, then appearing again, until they stabilise. This was arborisation alright, but in larval jellyfish.

In clinical neurology, the field I work in, there's a different emphasis on neuroplasticity. It's understood not as 'growth' or 'transformation', but as the mechanism of repair or compensation after devastating injury. The tone is different too, of course: more circumspect, less certain, at least as far as hard evidence goes. The effects of plasticity are seen clearly on brain scans taken at different intervals after a traumatic injury, for example, but its mechanisms, and the extent of its capacity to restore the injured brain, remain vague. At present there are no commonly used drug interventions with the power to significantly promote it. But it remains a term in daily use. Every trauma patient will, after they are sufficiently re-oriented in space and time, be given a basic lesson in the brain's ability to heal itself: that the restoration of lost speech, a paralysed leg, amnesia, an altered personality, depend on old pathways being restored, or compensatory pathways being forged. Some patients make complete recoveries, many don't. Most clinicians cite two years as the length of the window in which such changes might be seen in the adult brain. A few make it longer, three to five years; a few are more conservative, confining the window to eighteen months. In the absence of detailed evidence, this becomes a matter of convention rather than science. It's also clinically strategic, some-thing to give the patient and her family hope, the motivation to rehab, to keep emotional devastation at bay for as long as they remain 'in' the window. In this context plasticity is often more a matter of faith than science. It's also a way of distracting everyone (including the clinicians) from the terrible reality of how little can be done medically for the patients, how a significant percentage of their recovery remains in the lap of the gods. Then the window closes.

The PIS continued, 'In depression, connections between regions involved in cognitive, emotional and memory processing appear to be weaker, and the brain's ability to form new connections in these regions also seems to be reduced.'

Imperial is now widely known for psychedelic imaging in the field of mental health. Since the first wave of research there in the Fifties and Sixties, advanced neuroimaging technologies have been developed which can map the effects of psychedelics on specific areas of the brain.

One of them is the default mode network, a collection of structures in the mid-brain associated with mind-wandering, remembering the past and planning for the future – all those self-referring thoughts that demand their thinking. Another is the salience network: interconnected regions of the brain that select which stimuli are deserving of our attention. Some kind of dysregulation in these networks is thought to be associated with the experiences of 'meaninglessness', the negative appraisals of self and mental rigidity that are symptomatic of clinical depression.

To date, much of the neuroimaging research has depended on observing general changes in levels of activation across these networks following psychedelic treatment. Plasticity, which happens at the level of individual neurons, has been inferred rather than observed directly. The Imperial study I was being screened for aimed to take this a stage further. Combining fMRI imaging, which offers the precise location of activated areas, with PET scans, which allow any changes to be tracked across time, the intention was to observe synaptogenesis as it was happening. As per the PIS, 'There is a growing amount of evidence suggesting that ketamine's anti-depressant action may stem from temporary enhancement of neuroplasticity in important areas.'

A quick search of the literature suggested that most of the 'growing' evidence was indeed inferred rather than directly observed, based on changes to larger patterns of brain activation or to neurochemistry. The only other evidence comes from animals: a couple of studies reported growth of dendritic spines in rats. But there are limits to translating rodent psychiatry: however grim it might be to spend your entire, brief life confined to an environmentally impoverished cage, the rat cannot be meaningfully diagnosed with depression or other human mental health conditions.

The lack of direct evidence reflected something of how provisional and callow much of the neuropsychiatric research on psychedelic therapy was. Even so, the current study *was* ground-breaking, using state-of-the-art technology in a clinically relevant area of investigation, and at the 'hard science' end, compared to the vast majority of therapeutic research. It would also be eye-wateringly expensive: running scanners over multiple sessions with all the adjacent tests and staffing requirements meant the cost, even with a cohort of fewer than ten volunteers, would run into the high six figures.

The finance and science of psychedelic medicine are complexly entwined. In 2013 the US Food and Drug Administration (FDA) designated a variant of ketamine a 'breakthrough therapy' on account of its apparent ability to reverse the acute symptoms of 'treatment-resistant' depression. This led to a pharmaceutical arms race, the details of which were explained to me by Josh Hardman, founder of Psychedelic Alpha, one of the most reliable sources of financial information and commentary on the nascent psychedelic 'sector'. 'Ketamine has been used "off-label" for a number of years in the treatment of depression,' he told me, 'but according to the calculus and playbook of pharma there's little "defensibility" in these cases: it's not patented in any meaningful way for these uses, so there was no prospect of digging a meaningful IP moat around it.'

This changed in 2013 when the pharmaceutical company Janssen decided to use what Hardman called a 'textbook procedure from the pharma playbook' to bring a variant of ketamine to market with patent protection. It chose one variant, s-ketamine (or 'esketamine'), and partnered it with a specific drug-delivery mechanism, in this case a nasal spray. They then sought and achieved patent protection from the relevant government body on the intranasal administration of esketamine in treatment-resistant depression, under the trade name Spravato. In other words, certain design choices that had little to do with empirical evidence allowed them regulatory exclusivity on their variant and permitted them to market it as a 'new chemical entity'. This type of 'innovation', commonplace in the broader pharmaceutical industry, is, Hardman suggests, now entering the psychedelic sector.

But Janssen ran into significant problems with the health economics of its 'invention', as the price was forced up to $6,785 for a month of treatments twice a week. 'Remember,' explained Hardman, 'ketamine, unlike other psychedelic interventions, is associated with "temporary" changes in neuroplasticity; meaning that its prescription has a different economic model than "one-off" treatments.' Then, he went on, there was the fact that there was little long-term evidence for its efficacy. 'Some experts, like former FDA reviewer Erick Turner, were flagging that even the more short-term data showed only modest efficacy and raised some concerns over patient safety.' These factors have meant that even if Janssen is able to convince healthcare systems that it has a product with a novel mechanism of action, the cost and lack of evidence make it very difficult to produce a convincing economic case for a health authority, especially in the UK, where NICE (the National Institute for Health and Care Excellence) has a cost-effectiveness requirement for recommending treatments for the NHS. Meanwhile, Hardman told me, Canada had flat-out refused to grant Spravato data protection, the Canadian court finding that it did not warrant the designation of 'novel compound'.

Such limitations do not obtain in the US. The FDA's initial approval of esketamine involved loosening its definition of 'treatment-resistant depression'. Previously this diagnosis had been restricted to those who had tried two classes of anti-depressant medication (there are several, including SSRIs, SNRIs, tricyclics). It changed this to mean any two different pills: i.e. it could be the same class of anti-depressant, as long as over the course of their depression history the patient had taken two different brands. Given the whims of prescribers and patients, this sets a very low diagnostic threshold. Despite this, the initial FDA approval remains, and much of the subsequent clinical research has adopted the same criteria for treatment-resistant depression. Even with such low-hanging fruit, Hardman told me, Spravato has not quite been the blockbuster Janssen had hoped for, though it remains lucrative by most standards, just not by pharma standards: by 2029, Global Data predicts, it will generate world sales of approximately $383 million.

The Imperial study site was a Victorian hospital in Ladbroke Grove, its sandstone facade blackened by a century of grime. I had worked here before, in a job which consisted of diagnosing rarer forms of dementia. Back then, 'neuroplasticity' and 'psychedelic therapy' were not part of the vocabulary of such clinics. Many who take psychedelics report a temporary 'ego-death' – Imperial's metaphor for the apparent dissolution of selfhood – but for those dementia patients ego-death was permanent, the shattering of their identity being perhaps the hardest part, especially for the families. That had been fifteen years ago. In the last few years there have been a smattering of studies that point to the possible remedial effects of psychedelics in certain neurodegenerative conditions: might it be that fifteen years from now the dementing could trip their way back to health? It seems unlikely.

I recalled that the neighbouring building was a small Carmelite convent of white-robed nuns, an order that had dedicated itself to contemplative prayer for 800 years. Even now the sisters never leave the cloister, their lives the order's own informal experiment on ego-death, on default mode network regulation, on synaptogenesis conducted by an extraordinarily ascetic self-selecting cohort. As with psychedelics there was a trickle of evidence that monastic lifestyles – the simple diet, meditation, ongoing study – foster resilient brains and, as we will see, there are other convergences between these seemingly antipodean realms.

The psychedelic department was so new that the information desk hadn't even heard of it. After several internal calls I received directions, but still got lost, lack of signage not a good look for a hospital known for its dementia care. The team were based in what I remembered as the outpatient dining room. Two young men greeted me from behind laptops, one from Italy, the other a Northerner, both in jeans with ironed creases, college sports sweatshirts, ruggish, untameable hair and black-rimmed glasses; the uniform of the science PhD. The Principal Investigator appeared. He had been there since the centre was established, alongside the luminaries of the psychedelic new wave Professor David Nutt and Dr Robin Carhart-Harris. A few months previously Carhart-Harris had been successfully wooed to Neuroscape at University of California San Francisco Hospital in the Bay Area – Michael Pollan had been an early champion – by the promise of a 'different challenge', a professorship, bigger resources, more state-of-the-art tech. It was certainly a long way from this re-purposed dining room in West London.

I had asked for a meeting because I was curious, which is science for 'nervous'. I said I wanted to see the 'setting' for my first trip; really I wanted to find out what squirly strain of shit went down on ketamine at 'super-doses'. I was dressed uncharacteristically: cream chinos, a blue-striped shirt under a sports jacket; a uniform I wanted to say 'studious', 'responsible', 'humourless' and, most important of all, 'psychedelically naïve'. The Principal Investigator, meanwhile, was all in black: a man who over the last decade had witnessed so many clinically induced ego-deaths, I imagined, that he'd become his own cancellation or negative. He was friendly enough, if a little guarded, I assumed on account of the unusualness of candidates requesting a site visit, and answered my questions about dosage, biomarkers and scanning protocols matter-of-factly. This was the first experiment in the partnership between Imperial and this West London NHS trust, he told me, and such an innovative and ambitious study had only been made possible by a financial commitment from the manufacturers of the new scanner. But, he added, shoulders lowered under the weight of invisible concerns, the study's huge expense meant other collaborations might be necessary to get it over the line.

I steered the conversation towards my domain of interest: what that much ketamine might *feel* like. The PIS ran to thousands of words on the carefulness of the method, the screening, the logistics, the rationale, but not one on the colour and tone of the experience itself. 'I mean all that technical, bureaucratic talk obscuring the experience that lies in wait – it's a little Kafkaesque.'

'You're worried you'll wake up as a giant beetle?'

'No . . . should I be?'

'Nobody's hiding anything, Dr Mitchell.'

Which is exactly what the officials in Kafka say. At which point I asked the Principal Investigator directly: had he or either of his assistants ever taken ketamine?

They looked sheepish; none of them had. They weren't hiding; they just didn't know.

What formal experimental interest were they taking, I asked, in the participant's subjective experience?

There would be some questionnaires on personality traits ('openness' in particular), it appeared, a brief interview on the contents of the trip itself and pre- and post-debriefings by a clinical psychologist. Otherwise the study's area of interest was exclusively neuroplasticity. Experience appeared to be more of an afterthought, a side-effect even.

I wondered how Ethics could sanction something that was effectively a black hole. How could participants give 'informed consent' to a procedure so lacking in information? It suggested something about the disposition of science itself towards psychedelic experience, as though ketamine had retained its medical status as a dissociative anaesthetic; that nothing happened while you were under (beyond the sprouting of dendritic spines) until you woke to find something missing (a tumour) or something new (a hip) and crossed your fingers that things would be better than they were.

I had been drawn to clinical neuroscience by the strangeness of its phenomena – what you might indeed call their psychedelic quality – long before either my personal investigations with the drugs or the current resurgence of research interest. It began with stories: the Russian neuropsychologist Alexander Luria's account of the boy who was unable to forget the titles and order of the books he saw on the shelves above his schoolmaster's head forty years later; the British neurologist Oliver Sacks's accounts of the man who mistook his wife for a hat and patients awakened from comas by music; the Indian American neuroscientist V. S. Ramachandran sticking pins into phantom limbs; and the Portuguese American neuroscientist Antonio Damasio building his theories of cognition around the unfortunate fate of the railway worker Phineas Gage, who, in one of nature's crueller magic tricks, had a tamping iron fired straight through his head that left him totally intact apart from the dark revolution of his personality.

At its best neuroscience was profoundly curious about the experience of reality, its explanations enhancing strangeness rather than reducing it. The most consummate narrators and practitioners displayed an imaginative, literary-like engagement with the neurology of experience. (Oliver Sacks considered Luria's long case studies, *The Man with the Shattered World* and *The Mind of the Mnemonist*, 'neurological novels'.) As I would find out when I commenced my training, this ran counter to much of the clinical and academic reality, where such curiosity was relegated by the constraints of time (the short shrift of brief outpatient appointments), but also by an incuriously reductionist way of seeing. It is standard in these fields to think and talk of the brain as a computer, a 'meat machine', which leads to the quality of experience being described as if it were a form of computation: subjectivity, the lived experience of an individual's unique perspective, becomes a matter of 'predictive processing', of 'bits of information flow' that 'render three-dimensional representations' in the mind etc.

My doctoral research was on the subjective experience of anosognosia, from the Greek for 'not knowing'. The anosognosic patient loses an aspect of functioning, usually following a stroke; it might be the use of their left arm, or aspects of their memory or speech. With that specific loss comes another, the awareness of it. If the patient with limb paralysis is handed a tray, they are likely to receive it with their hand not in the tray's middle, like an amputee would, but at one side, expecting the injured arm to retain its normal functioning, with the result that tea would be spilt. Seeing it in the flesh has a strange, belief-defying quality to it; one imagines it's not going to happen, that the patient, whose cognition is otherwise intact, will make the necessary adjustments. Though it feels strange, it feels familiar too, as though something similar is happening in all of us in a more commonplace psychological way, as we keep on doing the same ill-advised things, even though we are quite aware we will end up with tea all down our shirt.

To explain this, most theories today congregate around 'disrupted feedback mechanisms', which have their basis in the model of the brain as an information-processing computer. Before this model became the norm, though, there had been a theory which considered it the Freudian defence mechanism of denial. My research aimed at finding a third way, with experiments designed to show that awareness was not simply present or absent in the way the theories suggested, but could be there in an unstable, partial, fluctuating way – which itself might depend on how the impairment was being referred to, scrutinised or talked about medically. In psychedelic terms you might say I was investigating whether anosognosia is dependent on *setting*.

I mention all this because it reflects my own bias towards bizarre phenomena in neuroscience, which long predates, though certainly intersects with, my current interest in psychedelics, and also because there was one patient in my research group whose presentation was so startlingly psychedelic (though at the time I would not have thought it so) that I have been unable to forget him. Rather the spectre of him grows in strength and significance the further I myself venture into psychedelic space.

During the information-gathering phase of my research, I would get a call from one or other neuro-rehabilitation unit in London whenever they admitted a potential candidate for my study. I was keen to visit the patients soon after their arrival; anosognosia was commonly a transient condition that remitted a few weeks after stroke or trauma. But this was not always the case. One morning I visited a small hospital in Greenwich to screen a recent arrival. Afterwards the ward clerk, who knew of my research, suggested I might want to visit another patient, a Moroccan doctor who'd suffered a right hemisphere stroke in his native country and been medically transferred to London. That was five years ago and, the clerk told me, little had changed since.

Off the painfully bright corridor with its cold linoleum floor was the warm, dark haze of Dr Aziz's lair. Bars of light streamed through a Venetian blind, enough for me to make out the delicate, lustrous patterns on the rug on which I now stood. The air was scented with saffron and recently consumed baklava, in stark contrast to the blast of detergent and urine from the corridor. Aziz, a huge man in his mid-fifties, lay on his side between two electric fans, dressed in a golden kaftan and a cotton skull cap. On his bedside table, a plate of dates and almonds, a pot of mint tea and the Qur'an.

Aziz told me that he currently resided in a respectable suburb of Marrakesh, that he was just taking a break after a busy morning in surgery, and would resume his list shortly. I should mention that though he had remained profoundly paralysed down one side of his body these last five years, his speech and intellectual functions remained largely intact. Both his sister and his medical records from Morocco reported no history of psychiatric problems prior to his stroke; he had been an equable bachelor with a successful city practice and a particular fondness for Persian artefacts and music.

I moved on to testing out his anosognosia. Would Dr Aziz be able to perform his normal manual duties this afternoon?

Naturally, he replied: as soon as he'd recharged and taken a light lunch.

Pressing him, I asked if he would he show me with mime how he might examine his patient's reflexes.

Dr Aziz declined, gazing at me as though it were the question of an idiot to an idiot, and reminding me that he was enjoying a well-earned rest.

I asked him if he'd mind if I examined his paralysed hand a moment. I scratched it lightly and asked him if he felt anything.

'That's strange, I don't . . . Very strange – it was working perfectly a moment ago . . .' Aziz spoke English like a lightly orientalised English duke. 'I'm sure if you let it rest a moment it will be right as rain.'

I asked him to close his eyes while I touched one of his fingers: all he had to do was tell me which one. He obliged, while obviously again considering it an imbecilically simple task right up until the moment he was unable to execute it.

'Yikes . . . That *is* very peculiar!'

But, as before, he seemed only momentarily concerned. With that Aziz checked his watch and asked if I wouldn't mind coming back at a later date: he had a patient booked in for 1 p.m., for the removal of an awkwardly located boil, and he would need to settle himself first.

Stepping back out into the shockingly bright corridor was like shifting dimensions. I left Aziz's room less with the sense that he was hallucinating, than that I had hallucinated him, such was the coherence of his presentation, and the hyperrealism of its staging. His anosognosia was complex, certainly, beyond either of the simplistic formulations of 'denial' or 'failures of feedback'. Apart from those brief moments where I was moving or touching his hand, Aziz's reality overpowered mine. Beyond the facts themselves, I had just stepped out of a GP's apartment somewhere in Marrakesh, as he lunched between patients.

While short on the details of psychedelic experience, the PIS was clearly proud of its state-of-the art neuroimaging protocols for high dose ketamine and DMT. Neuroscience has long been criticised for its simplistic over-attachment to neuroimaging, and the inflated inferences drawn from it. Several decades' worth of enhanced technologies and novel protocols have attempted to locate the brain activity that relates to specific emotions and behaviours: 'anger', 'jealousy', 'the munchies' – even 'the area associated with the response to a remotely controlled charging bull'. But many such studies, it is argued, provide more of a cultural X-ray of our current preferences and obsessions than a scientific explanation of the workings of the brain. Some of the criticism is narrowly methodological, focusing on the limits of what can actually be inferred from particular imaging technologies or the artificial nature of the tasks subjects are made to perform. Some of it is more broadly philosophical: that the whole project is an exercise in 'neo-phrenology', the modern equivalent of the Victorian fancy of reading that odd bump on the side of your head as the idiosyncratic signature of 'amantiveness' – love, in other words; or that suggesting that certain areas of the brain 'speak' or 'lie' or 'like cheese and onion crisps' ignores the fact that the brain does not somehow act independently of an embodied person. After all, everything we think and feel will manifest in the brain, so finding evidence of activity in the brain doesn't mean we have somehow located the 'origin' of those thoughts or feelings.

Though still in its formative stages, neuroscientific psychedelic research has, according to some commentators, revealed similar tendencies, assuming that mental events are localisable; that they map uniquely to dedicated brain circuits; and (perhaps most saliently, in the case of psychedelics, where setting is crucial) that they are independent of the larger context.[2] So while, for example, the idea that the brain contains a self, or homunculus, that integrates all sensory and perceptual inputs has long been disregarded as folk intuition, a fictive 'ghost in the machine', there are, as one eminent neurologist astutely notes, ghosts of ghosts in the machine, evident in the current obsession with default mode and salience networks. The current focus on such networks suggests we still believe that a more provisional, more qualified version of selfhood resides there; that that's where a person's essence is to be found – which once again overshadows any real interest in the actual experiences of people.

With some key exceptions – Professor Anil Seth's research into consciousness, for example, which is deeply curious about the experience of reality, which he calls 'a controlled hallucination' – there is a relative lack of academic interest in the experiences themselves. In part this reflects a significant reduction over the last few years in public and philanthropical investment in psychedelic research, and a concomitant increase in private or commercial funding. According to Josh Hardman, such investors prefer to back therapeutic-based research where there are opportunities for patenting and therefore profitability rather than more investigative neuroscience. And this disinterest tells us something important about the psychedelic future being prepared for us.

'Psychoplastogens' (aka pseudodelics) are a new trend in drug development. They are synthetic versions of existing psychedelic drugs that supposedly have the capacity to promote synaptogenesis or neuroplasticity, but without the perceptual distortions, hallucinations and other subjective derangements of the trip. In other words, the defining characteristic of the psychedelic – that which terrifies, weirdifies or sacralises – is erased. Hardman explained the rationale: given the cost of monitoring an hours-long psychedelic trip, the fears many hold about taking them (clinically justified or otherwise), and the challenge of integrating this treatment model into the prevailing healthcare system,[3] some companies have sought to do away with the trip altogether.

The investment is not on a casual scale: psychoplastogen drug discovery and development outfit Delix Therapeutics closed a $70 million Series A financing deal in 2021, and Professor Bryan Roth at UNC got $26.9 million from an agency specialising in developing breakthrough technologies for national security (Defense Advanced Research Projects Agency – DARPA). Hardman explained that beyond the potential cost savings of developing take-home psychoplastogens like this, 'they might also lend themselves to chronic conditions that may require more frequent dosing, like pain-related conditions.' They might also prove suitable for patients for whom psychedelic therapy would be inappropriate (those with a diagnosis of schizophrenia or bipolar, for example). The research is extremely preliminary, even by the standards of general psychedelic research: 'unless some of the researchers involved have tasted their research materials, we're still at the stage of rat models here,' Hardman said at the time of writing. 'It's yet to be seen whether these drugs "work" in humans, and even whether they're actually non-hallucinogenic.' But the trend – away from the experience of psychedelics and towards neuroplasticity – is significant. It points

toward the familiar aims of conventional psychiatry: a patient pas-
sively takes a pill, hoping that their brain changes its chemistry
while every other aspect of reality remains the same, for better or
for worse.

Back at the hospital in West London I'm making a nuisance of myself,
wondering out loud how the researchers can hope to contain the
possible dangers of taking 'super-doses' of an under-researched psychedelic drug. 'A drug none of you claim personal familiarity with,
which is fed in extravagant doses, inter-arterially, deliberately select-
ing inexperienced participants who have to navigate other-worldly
realms while being trapped in a horrible, raucous metal coffin!'

'No, no, no!' they chortled. 'You must have misunderstood.'

The treatment itself, the Principal Investigator told me, his
expression lightening, would be in a 'specially curated suite'. And
with a new bounce in his step he beckoned me towards an adjacent
room. He pointed out the artwork they'd specially commissioned.
On the wall opposite each bed were two painted golden whorls of
undulating concentric circles, fragmenting into large gold blobs that
crawled up the wall and across the ceiling, shrinking as they reached
the doorway into beads of golden mercury, like fungal spores, like . . .
I felt the pressure of approbation a parent feels when their child
shows them a painting.

The team at Imperial had come a long way in a decade, the
Principal Investigator was explaining: forty strong, highly multidisciplinary . . . they had commissioned designers and composers
to maximise the setting's potential to 'positively mobilise' the psychedelic experience.

He was still waiting for me to say something.

'It's reminds me of . . . airborne cappuccino.'

My site visit was over. With no formal information about what I
could expect to experience on super doses of ketamine my research
was forced underground. Panda (her real name) had worked with
the Multidisciplinary Association for Psychedelic Studies (MAPS) in

the US for years organising their MDMA trials, then left to set up her own integration service for psychedelic users in LA County. If you were planning a trip, then Panda's therapists could prepare you and help make sense of the experience on the other end. It was just the illegal bit in the middle where you were on your own. Business was booming: she had six full-time clinicians and plans to build an 'integration farm' in the pastureland of the San Bernardino.

When she was not busy being one of an increasing number of psychedelic entrepreneurs, Panda was a committed first-person researcher. 'Ketamine is like a video game,' she told me. There were numerous levels – fifteen, twenty, even – according to slight increments in dosage. Panda's favourite was 'Fur'; at this level one's field of vision appears entirely coated in animal pelt – chairs, computers, faces, fruit, which sway like a field of wheat when blown upon, and feel kitten-soft when touched. 'It's so fricking cute.'

It sounded terrifying.

Then there was the Level of the Crystal Waterfall, where the entire world is seen through a veil of frozen water, sparkling like diamanté, giving the impression of cascading or coursing without actually moving. To her it was 'enchanting, the sort of place you'd want to live in forever'. But it reminded me of akinetopsia, a rare neurological condition caused by specific strokes or brain injuries, in which people retain perfect visual acuity but lose the ability to perceive motion – the world turned to time-lapse photography with big jumps – which can also last forever.

Panda's video-game levels sounded so tailored to her, to her 'set', so quirky, so absolute, so black and white, like a Panda. Surely a middle-aged man's set would yield a different kind of 'game': the Level of the Failure to Invest Earlier, the Level of Prostate Anxiety?

Apparently not. These were invariant, psychophysical progressions, Panda explained, like the numerous stages of 'insight' one passes through on a Vipassana meditation retreat: everyone followed the same map of the same territory, no matter what they preferred or believed.

I found *that* hard to believe.

And then there was the video game's secret level: the fabled K-hole. Depending who you asked, it was a state of absolute dissociation or total oneness with everything; perfect serenity or wholesale terror; the holy grail or to be avoided at all costs. It was, I thought, an apt metaphor for where I was, where we all were: on the brink of a pill that delivered the unknowable, and it was up to the consumer whether this was the best or worst experience of their life, if indeed they were one and the same thing.

A week after my meeting with the team at Imperial I received a short email. A new collaboration with a 'private interest' had necessitated changes to the study's eligibility criteria. Volunteers were now required to be completely 'psychedelically naïve'. My having taken ayahuasca with Aurora disqualified me.

I called up the centre and got the Italian PhD student. Was it all the questions? The remark about the coffee?

There was nothing to be done. I had worked for years in healthcare, witnessed countless desperate patients fail to get on trials for new therapies. Now I knew what it was like.

Psychedelically naïve. Emerging research suggests that, beyond a well-designed space, the active ingredients of 'setting', which has such a determining effect on one's psychedelic experience, extend to include one's entire cultural context. If so, the hype was bound to have significant effects: positive expectations among self-selecting psychedelically curious participants which inevitably contribute to the favourable outcomes reported by clinical trials. And unless you were the equivalent of the nuns, cloistered from every headline of social media, dedicating your entire life to training the mind to undo every unholy expectation as soon as it arose, true psychedelic naïvety was impossible. And yet a different kind of naïvety infuses the entire project. Consciously or otherwise, the academic/corporate mission to promote psychedelics involves presenting them as a 'novel medical intervention' that owes nothing to the pre-existing, never-went-away, underground tradition. So naïvety as a strategy to split itself off from origins, to erase unwanted history.

At the practical level – from the perspective of the potentially extreme nature of high-dose ketamine – would it not be wiser, safer, more ethical, to have volunteers who had spent hundreds of hours exploring the terrifying trapdoors – the demons as well as the angels – of psychedelic space? Recent research indicated that a significant portion of psychedelic trialists had difficult experiences, with a percentage of those experiencing major depression and suicidality (of which more later). From that perspective it made sense to test these dangerous, super-charged vehicles on expert drivers rather than beginners.

Despite a dizzying number of studies underway, 422 as I write, many of them are exploratory, under-powered, and lacking the necessary methical rigour. The study I had failed to participate in had nine participants. That means that one outlier would change the results significantly; two would constitute over 20 per cent of the sample. It would clearly lack the power to make robust claims about neuroplasticity in the general population. There was something wilfully naïve about that, too.

Of course, those who work in psychedelic science – as in any area of science – are not selected at random, and like all scientists they bring with them their personal motivations, their intentions, their *sets*. Researcher Manoj Doss, part of the team at Johns Hopkins, stated that he knew of only one colleague who hadn't tried psychedelics at some point.[4] With an experience as strange, esoteric, and potentially terrifying as psychedelics it may be useful to have researchers or clinicians who are the opposite of naïve, who've been there before. One New Age nostrum has it that the journeyer can only go as far as the guide has been. But the trouble is that a transformationally positive personal experience for the researcher can lead to unconscious biases at any stage of the scientific investigation and its publication. 'All the major papers that've been published in the last five to eight years I've reviewed – all of them' (the words of the Imperial College London-affiliated psychiatrist Ben Sessa). 'I approved them all. I mean, I suppose maybe I should be less biased. But I approved them all. I think they're all great papers . . . '[5]

It's been noted how researchers commonly have potentially conflicting interests as consultants for pharma companies, both in the
US, where such straddling is the norm in pharmacology, and in the
UK, where it isn't. Hardman described congratulating a researcher
on their appointment to an advisory position at a psychedelic start-
up. 'For a moment he looked confused, unable to recall any such
appointment for a few seconds, until I jogged his memory with his
photo and pictures of his institutional logos which featured promi-
nently on the company's pitch deck.' The Principal Investigator in
the ketamine study was an advisor for Clerkenwell Health, a company which looks 'to support the commercial research ecosystem'
in support of 'psychedelic companies flourishing in Europe'.

Rejected by Imperial, I approached another well-known psychedelic researcher at Greenwich University, and he was able to
recommend me for a trial, not in his own department, but with
Small Pharma, a biotech company whose specialty is developing
various short-acting psychedelics for mental health. (Sadly, I wasn't
depressed enough to meet criteria.)

Meanwhile, the vast majority of psychedelic research has fallen
short of the gold standard of the double-blind controlled trial. This
is in part a function of economics – organising such trials comes at
huge expense. But blinding represents a profound conundrum in
this field in particular: how to mask or blind a group of volunteers
when people tend to know when they're tripping balls? Or when
ethics committees generally request a description of some of the
anticipated experience to ensure the consent is *informed*? Most
research to date doesn't measure its patients' expectations let alone
attempt to reduce them.

I had my own expectations too, of course.

Like most of my colleagues in medical science I was a more or
less explicit 'materialist' or 'physicalist' – believing that all the phenomena of human experience can be understood as the expression
of physical occurrences in the brain. At different times I had enter-
tained other possibilities, but if you put me against a hospital wall
and held a neurosurgical laser to my head, I'd repeat back the

fundamentals of 'materialism': the mind and brain are one, outside of it there is something called reality which is knowable scientifically and conforms to laws that have emerged from that science.

At first I extended this underlying belief to my experiment with psychedelics. Before I first drank ayahuasca I would have coarsely predicted that my experiences would follow the logic of that materialist perspective: I mess with my brain in a specific way and thus specific things happen to me – however strange – and stop happening when the chemicals disappear. From that vantage point, psychedelics tell us nothing about the nature of reality. That stays the same; it would be my brain that was changing.

For many, psychedelic experiences do lead to a shift in one's overall view of reality: the 'old' self is somehow obliterated, a valve opens on perception, something more true, more 'real' is revealed.[6] But to a clinical neuroscientist, such changes are commonplace – as in Alzheimer's or closed head injuries, for example – and such effects have been investigated academically for decades. The mere fact of having those experiences doesn't make their insights valid, is what I would have said, with irritating knowingness.

But I say more or less materialist, because I also carried a sense that the whole perspective of neuroscience can be overly reductive, that its incuriosity about first-person subjective experience meant it was missing something. Phenomenology is a philosophical tradition that has called into question the insistence on a single, independent, wholly determinable, objective reality. It argues not only that our subjective experiences shape our perception of reality but also that any meaningful or useful notion of reality has to take into account the unavoidable fact that it can only be encountered via experience. It's enjoyed something of its own renaissance in recent years with writers like Iain McGilchrist in the field of neurology, David Abram in ecology and Jonathan Bate in literary criticism applying some of its implications to their various arguments. Even sworn materialists like Anil Seth have found certain of its ideas useful to their way of thinking.

Science recognises the need for 'reflexivity' – taking into account the effect of its observation – in certain domains like quantum mechanics, or in the inevitability of the placebo effect, but most science assumes that our observations tell us something reliable about the world out there, independent of whoever is doing the observing. Phenomenology, by contrast, aligns itself with the view that there is nothing objectively available outside that isn't a construct of our minds. By the terms of materialism this position is difficult to justify. But in its curiosity about 'phenomena' – meaning that which is experienced, felt, seen – it offers a rich vantage point from which to understand what it's like to be an embodied person, how strange even the most familiar environments really are, or what taking a trip might suggest about the nature of reality.

Attempting to make sense of Dr Aziz, the bizarre case of anosognosia I encountered in my doctoral research, I turned to the work of the French philosopher Maurice Merleau-Ponty, a key proponent of phenomenology. For Merleau-Ponty it is only via our bodies and the perceptions afforded by our bodies – by having intention, movement and feeling – that we are able to construct a notion of the world and a notion of ourselves; something we are continuously in the process of doing. When that process goes well, we have 'maximum grip' on our reality, meaning our intentions are translated into the outcomes we predict. But when it is impaired, then that capacity of the self is lost and the patient must abandon their reality, forced to construct a new notion of the world out of their new realm of perceptions, even while they may struggle to break free of their previous one. This has some intuititve force when I think of my encounter with Aziz. Merleau-Ponty used 'breakdown' or 'edge' cases – medical conditions like anosognosia, other extreme experiences, and certain movements in art – to ground his theory. But these edge cases reveal something that is part of the normal background of perception: confusion, instability, indeterminacy. We resolve this, he argued, by finding ways to break free of our habitual modes of attention and thereby attend more openly and directly to our physical and extraneous environments, or, as he put it, by finding the right 'body set' in relation to the world-out-there, allowing it to be encountered in all its strangeness.

Merleau-Ponty was a *first-person* philosophiser after my own heart; his experiments with pharmaceutical mescaline gave him another kind of 'breakdown experience' that informed his thinking. What I found seductive and poetic in Merleau-Ponty was the idea that the drug only illuminates what is obscured during (and by) normal perception: that we are in a mysterious relationship with an unspeakably strange world. Habit and convention entrench (or 'canalise', to use Carhart-Harris's latest metaphor) our attention, limit our neuroplasticity in contemporary terms, deaden our relationship to our bodies, and make us prisoners of our expectations, as stuck as Dr Aziz was in his mode of reality, and unlikely to move beyond it. Psychedelics offer us the potential to unstick ourselves.

At least, they do if you meet the criteria for a trial. 'Too naïve', 'not naïve enough', or just too plain 'healthy', which is not a term often levelled at me. I tried and failed to get on several more. My first journey was not only trip-less, therefore, but also officially drug-less, medical science robbing me of the entire 'phenomenology' of super-dose ketamine *and* its neuroplastic effects. There was not even a placebo; only dashed expectations. Hard not to take it personally, that this was karmic revenge for taking leave of my clinical responsibilities and heading off on my psychedelic tour.

So much for 'legal' scoring and clinically supervised tripping. As with twenty years earlier, it forced my drug-taking underground. (One final reminder to the prohibitionists: many of my colleagues have at one time or another experimented with psychedelic drugs. In the last few years, with more information about their therapeutic promise, this number has grown exponentially. In my opinion this in no way hinders their sober ability to treat your illness or hear what's on your mind.) Via WhatsApp, my friend Ben's daughter Izza ordered me enough ketamine for a super-dose. There would be no arterial line, no neuroimaging, no commissioned art or music. The cost: a little over £30 as opposed to low six figures. Izza would be trip-sitter, 'guide', co-consumer, DJ and integrator. She was in her mid-twenties and used ketamine on the back end of 'big' nights to wind down. This was the first time she'd be taking it 'straight' (in the morning, after a night's rest), 'non-recreationally', and in her father's kitchen in North London. I say 'non-recreationally' because I remained rather neurotically attached to framing this experience as a clinical 'experiment' so as not to revisit a history of abusing other drugs.

A little after 10 a.m. I 'insufflated' – science for 'snorted' – the first of three large mounds of fine white powder through a rolled ten-pound note and waited for my experience. Unlike the plant-based psychedelics it came with no ceremonial history to contend with, no remote cultural anthropology to study or ignore and, expelled from the research cohort, no doctors or scientific machinery. It felt liberating. The kitchen setting felt post-scientific, post-spiritual, post-modern. Izza put on Brian Eno.

On the other hand, my neurosis was spiking: the white lines and rolled notes, the grey light of morning in a strange house, brought back all those fraught, sordid hours lost to addiction in my early twenties. (Izza followed Eno with the Orb's 'Perpetual Dawn' – a staple from those days.) This was the opposite of liberating.

Several months later, I can say that super-dose ketamine was not the astonishing, auspicious beginning of my global psychedelic odyssey I had hoped for. For all the talk of French phenomenology, the mystery of the body, the secret life of the object, the trip itself was brief, mainly banal, disappointingly unrevelatory and, compared to the trips that lay ahead, hardly worthy of report.

For example:

'Ben ben ben ben ben ben ben . . .'

'What?' said Izza.

'Ben, like your dad's name, but each one sounds different . . . ben ben ben.'

'Not to me.' I was more than twice her age, and less than half as cool.

'Some stroke patients can only say one word . . . but they put everything they're thinking and feeling into it. *BEN . . . BEN . . . BEN! . . .* Try it.'

'No, I don't want to.'

I talked too much, like cocaine-fuelled self-talk, only wonkier, stupider: ketamine the pantomime-horse tranquilliser.

A couple of things stood out above the fray. First, taking psychedelics is like lifting the hood on the brain: you see something about the mechanics of consciousness – language, sensation, perception – in close-up or from an unfamiliar angle. As neuroscience becomes more and more part of the mainstream vernacular, many of us are curious about understanding ourselves in this way, which is part of the psychedelic attraction. Ketamine gave me a brief, skewed glimpse of the workings of my mind.

One finding that would prove consistent across all my psychedelic experiences is that the trip always begins with fear. Whatever my intention, however relaxed or peaceful I feel hours before, I feel a flood of cortisol or adrenaline washing through me within minutes of consuming, and before I know it I've turned into the neurochemical equivalent of a Navy SEAL: in my mind alone; primed to engage the enemy or bolt for the nearest exit. For me, taking a psychedelic drug means first learning how to down weapons, relax, remember that the war ended a long time ago.

The second thing was more abstract. Back in my friend's kitchen I mused on how the thirty years that separated Izza and me would inevitably weight our experiences differently. This led me down an extravagantly deep rabbit hole. Sixty years on from the first psychedelic wave, the same drugs now find themselves in an almost unrecognisable setting. At the time of my ketamine trip I had just finished reading Nobel Laureate Annie Ernaux's incredible memoir *The Years*,[7] which more or less covers those same six decades, and the book infiltrated my set that morning. In it, she describes how accelerating innovations in technology have made our memories ineradicable: social media, which appeared in my forties and Izza's teens, ensures they live forever, removing the depth of time, suspending us in an infinite present. We have 'drained reality', Ernaux tells us. And what, she asks, is the net effect of this on our imaginations? (And why – I'd like to ask – with all this *progress* is our mental health actually worsening?) The obscurity of previous ages, that fundament of personal privacy, has disappeared forever. Therefore, Ernaux states, we need new forms of obscurity, new ways of securing ourselves. Were psychedelics such ways, I wondered – the ultimate in non-disclosure agreements? Watching Izza effortlessly sync music and visuals on her iPad I had no idea what the world meant to her, and no way of telling her what it meant to me: how this state-of-the-art Islington kitchen, with its smart water purifier, its digital air fryer and phone-controlled coffee maker, resembled a space station next to the one I grew up in. I felt alone, out of time. The things around us did not last long enough to grow old alongside us. Suddenly I had acute, overwhelming kitchen nostalgia.

Ketamine was relatively new; white and manufactured like the appliances around us. If there was neuroplasticity, if there were dendrites sprouting, I thought, they were synthetic ones, plastic only in the way a fake Christmas tree is plastic. I felt powerful longing and regret: how could I have neglected to feed fridges when they'd been there all through my life when I'd needed them: silent, stolid, uncomplaining . . . ? I thought about placing my head in the washing machine and turning it on, to replicate the MRI.

There were no obvious video-game levels in the manner Panda had described; no fur other than where one would want fur – except perhaps at one point. After the last of the lines I had retreated to the toilet and sat there, according to Izza, for more than an hour, unable to send the right commands to my body to pee, and trying to call my sponsor Lenny for an impromptu 'meeting'. Eventually I got to my feet and looked back at the water tumbling over itself incessantly – the flush must have jammed: a vortex of looping swirls, just like the 'coffee design' on the walls of the treatment room at Imperial. And I saw that my mind was frozen at the bottom: my K-hole, recycled by the water's currents for eternity, which meant this might also be the fabled Level of the Crystal Waterfall. Panda and that Imperial swirls artist knew what they were on about after all.

All told, I found ketamine more shallow than it was deep, a private weirdness to insulate one from the weirdness of the world, the opposite of the enhanced 'connectedness' which is so often a finding in the therapeutic trials. But the following morning I detected traces of synaptogenesis: I didn't want my regular coffee, or to check Arsenal's transfer news. The idea of scrolling through my phone before getting out of bed had become what it was: helpless and gluttonous at the same time. The same held true the next day, and the day after that. Then the old habits slowly woke and took over again; three days of liquid, plastic freedom before it set like concrete once more.

2

Mystical Experiences

They were everywhere. Benedict Cumberbatch as Marvel superhero Dr Strange was putting paid to a bright green, fifty-storey Cyclopic octopus in Manhattan with trademark aristocratic insouciance. The doctor kept falling through inter-dimensional rabbit holes, literally. In one of these dimensions loud-coloured Blobs, their language the sound of squelchy paint, appeared to want to make inter-species love to Strange and his young companion. Nobody had actually seen these things before, *and* they were psychedelic clichés.

I pressed the unresponsive screen on the back of the seat in front of me, my fingers sticky with butter, the only healthy item on the in-flight meal. If the aesthetic inspiration for *Dr Strange in the Multiverse* was LSD (in an interview the film's director had spoken of his nerdish fidelity to Sixties comic art creator Steve Ditko), then *The Northman*, in which Nicole Kidman and Alexander Skarsgård mope their way through two-and-a-half hours of dreamlike, 'well-researched' Viking anthropology in muted palettes – shamanism, occult curses, endemic facial hair, pagan sex, ominous crows, and multiple disembowelments – had henbane, the legendary Scandinavian psychedelic plant, for its muse. Two films: one tragic, slow, atavistic, ecological; the other comic, cartoonish, quantum, esoteric. Both saturated with hallucinogens.

I was somewhere over Iceland, where thankfully disembowellings have abated these days, bound for California. My second trip would be the treatment of a patient with a diagnosis of intractable depression using therapy-assisted magic mushrooms, with myself in the role of therapist.

Though statistically many more people are either neutral or

negative about psychedelic usage than are positive, half of the sample population hold a positive view of psychedelic-assisted therapy. At the time of my trip such therapy remained illegal. But the patient in this case, Father Bede Healey, a Catholic monk slash psychoanalyst slash clinical psychologist, was desperate. He'd battled with clinical depression for most of his adult life. Over the years he'd been tried on numerous different classes and combinations of psychoactive drugs by a long line of psychiatrists. He'd sampled most of the mainstream psychotherapies and a few outré alternatives. In the last couple of years he'd smelt the psychedelic hype, hard to avoid where he lived in Berkeley, where the tripping never really went away and the university had its own estimable Center for the Science of Psychedelics. Bede's associate teaching position at the university allowed him to skim the research in scientific journals, and when I told him of my project in one of our regular phone conversations – we provided each other with informal peer supervision on difficult cases – he immediately asked me if I would consider being his psilocybin therapist.

As the flight droned on I kept flicking between movies, trying to find the non-psychedelic bits. The people either side of me preferred to watch their phones. Big-screen entertainment meant holding the devices inches from their faces and listening through high-end noise-cancelling headphones rather than the bits of string the airline provided. Immersion was what we were all after. That's what psychedelics at high doses guaranteed: no escaping their immediacy, their immanence. As with a plane journey, once you were on board you surrendered all control, going wherever it decided to take you, as though it had swallowed *you*. But imagine not this state-of-the-art Dreamliner with its non-naïve passengers, but one of the early biplane versions made out of canvas and twine, where ecstasy, speed, death, vomiting still meant something psychologically, because the experience of flying at unprecedented speeds thousands of feet above the earth's surface hadn't yet been completely habituated. Part of us longed for old-style planes: psychedelics, in other words.

As if on cue the aircraft jolted aggressively, the seatbelt sign pinged. The captain made a grave-voiced announcement about severe turbulence owing 'to sudden changes in . . .' Unconcerned, I put my noise-cancelling headphones back on before he could finish and tried to fall asleep. We didn't escape reality; we just became too familiar with it, over-learned it, underestimated it. Psychedelics, however, were de-familiarisers, capable of rebooting in us a sense of what the poet Gerard Manley Hopkins called 'the deep-down fresh-ness of things'. Only, now that they were everywhere, we were getting used to that too. One day soon we would need to create new mind-manifesting drugs to break through our habituation to psychedelics.

Dulces Sueños, the Airbnb apartment, was at Pine Mountain Lake, Toulume County, a gated community of 2,796 wealthy weekenders and retirees a stone's throw from Yosemite and a thirty-five-minute drive to Sonora Regional, the nearest place with urgent care. There was a local clinic four miles away, which I imagined might have a low-spec, portable defibrillator, but it was closed at weekends. The veterinary clinic was open on Saturdays, but they would be unwill-ing to treat us, even in an emergency, unless we rented a horse from the equestrian centre six miles away. The bottom line was we were a long way away from resus, and that was a little unsettling.

Though the clinical research on psilocybin-assisted therapy indi-cated it was medically safe, the numbers were small. Beyond the medical literature, counter-narratives were beginning to surface, like the *New York Times Magazine*'s podcast *Power Trip*, which detailed the death of a man in his seventies following a 'heroic' dose of mushrooms (5 grams plus), with a possible undetected co-morbidity. Bede had just turned seventy.

We discussed the risk from different angles. His doctor had given the 'all clear' at a recent check-up. His friend Frank, a retired psych-ology professor also in his seventies who had come with us, was a *shroom evangelist*. Frank had made several recent journeys, emer-ging from his experiences in 'better than ever' health. Bede was an experienced clinician, a psychodynamic psychotherapist who had been trained to reflect on his personal motivations, to guard against impulse, fantasy, idealism. Over many years, his depression had made a good portion of his life difficult to bear. Boosterism apart, the medical literature (and Frank) held out the hope of radical, enduring and near-instantaneous transformation. It was worth a try. But some of the responsibility and risk was mine, and Bede left that part for me to work out.

Dulces Sueños, Sweet Dreams. It was accidental, but the name was fitting. Every one of us makes several ineffably strange, phantasmagoric journeys every night of our lives. A little over a hundred years ago dreams were the 'new technology', Freud's royal road to the unconscious, enabling clinicians to treat that era's mental health epidemic (for 'depression' and 'anxiety disorders' read 'melancholia' and 'hysteria'). A century later dreams were, for the most part, therapeutically ignored. Maybe we were dreaming less, the capacity of our minds reduced by constant digital overload? Or were we just more dissociated from the dreams we had, requiring therefore stronger dreams in chemical form? Were psychedelics dream substitutes? In which case, were we being sold what we already had? Or were they simply a recognition that we needed regular doses of the irrational and fantastical; that these qualities are in some sense therapeutic? Were we so preoccupied with keeping them at bay, with being 'professionally reasonable' in a world that increasingly didn't make sense, that our minds had become bent out of shape?

Neuroscientific research suggested that psychedelics produce mental imagery similar to those experienced in dreaming, and a comparable loss of self-boundaries and cognitive control. Both depend on the activation of the same serotonin receptor sites in the brain. Contemporary neuroscience understands dreams in the terms of modern technology: they are 'virtual realities', simulated by the brain to promote the learning and organising of information and enable the implementation of tasks during wakefulness. Whatever personal idea we have about our dreamlife, under research conditions the evidence points to it being biased towards negative emotions: anxiety, fear, aggression. Half a century ago the British psychoanalyst Wilfred Bion suggested that one reason we sleep is to make possible difficult experiences that we cannot have while awake.[1] Similarly, modern neuroscientists speculate that the negative bias of dreams might point to part of their function: to support our capacity to tolerate or regulate challenging emotions. And yet 'dreams' in the vernacular sense remain synonymous with idealised, airbrushed hope.

There's an analogy with psychedelics. Despite the medical literature's signalling of therapeutic change through positive, even mystical experiences, it's a less well-documented certainty that many psychedelic experiences contain dark, challenging encounters. The clinical literature suggests that when such difficulties happen in assisted therapy, they are 'contained' by the therapist. Bion compared therapeutic containment to a mother 'who bears and absorbs her infant's emotional states, transforms them, and "interprets" them for him'. In psychedelic therapy this is done during the journey itself, in a process known technically as 'scaffolding', more or less the same reassurance one would give a child. It then continues more consciously during post-treatment integration sessions, which can help transform difficult or enigmatic material into useful opportunities for learning and insight. Neuroscience understands this in terms of synaptogenesis: the integration sessions typically take place during the 'afterglow' shortly after the journey, when patients may have increased brain plasticity, making them more open to new ways of thinking and acting.

But despite the positive framing, it's not always so easy to contain 'challenging experiences', as a recent article by Rachael Peterson in the *Harvard Bulletin* makes clear. Peterson describes her participation in the much-publicised Johns Hopkins 2018 trial of high-dose psilocybin for major depression. During her first session she had a transformative 'mystical experience', which represented a significant therapeutic breakthrough in the treatment of her depression. Her experience formed part of the published data, contributing to the statistical significance of the study's positive findings. Yet a week later, on the same dose with the same therapists in the exact same conditions, she felt 'terrifying, bottomless panic', an experience she has come to think of as 'anti-therapeutic', 'a religious experience in reverse'. I will consider the nature and implications of bad trips in the next chapter, but for now what's important is that the researchers made no attempt to report the details or even the fact of her second experience. 'Amid growing hype that psychedelics are a panacea for mental illness, I worry,' writes Peterson: 'is my partial testimony being co-opted to support a medicalisation effort I increasingly doubt can fully attend to the weird wildness of these medicines?'

I continued to deliberate on my responsibility to Bede. That there was a degree of psychological risk was undeniable. But my patient was clinically trained to be reflective, understood better than I did that risk, rather than being something to avoid automatically, can, in therapy as in life, prove the crucial ingredient; that challenge, intensity, discomfort, loss of control, might be just what the patient needs, and especially a patient who has, over decades, been slowly turned to stone by the flatness and un-responsivity that characterise chronic depression.

Please do check to see if refuse bags are already in the cans, they cost 10$ each, says the 'Welcome to Dulces Sueños File'. *Plumping cushions repeatedly each day is one good way of ensuring their longevity . . .* So controlling, so aggressive. This holiday home had the feel of a labour camp. If only they knew . . . And the irony of it, because treating it as our home – specifically, treating the 'living room' as 'our living room' – was a central tenet in the literature of psychedelic-assisted therapy.

Over the past twenty years, the living-room-like setting at Johns Hopkins has offered a model for psychedelic research facilities at other universities, to the extent that for psychedelic retreat centres, underground therapies, handbooks and guides, 'living-room-like' has, according to Tehseen Noorani (a former Science and Technology Studies post-doctoral student at Johns Hopkins who was invited to observe the research), become a marker of competence when administering psychedelic therapy. Only, these 'living rooms' are constructed, literally and metaphorically, by the scientists not just to locate the patient in a supposedly homely setting, but also to remove them from their actual situation: the middle of a hospital complex, surrounded by all that pain and suffering. Noorani describes them as 'a series of inter-connected office rooms . . . carefully overhauled to have lamps with soft lighting, colourful paintings (including the German Expressionist Franz Marc's *Tirol*) adorning the walls, flowers, cushions and rugs with textures and patterns, bunting, a sofa with pillows and blankets for participants to lie on and a large bookshelf full of picture books on art and nature'.[2]

Timothy Leary, the Harvard psychologist who did much to advance psilocybin research in the Sixties before going famously 'off-road', described the psychedelic setting as the room's 'atmosphere, its weather'. But this feels too naturalistic a metaphor. In Noorani's description, the living room starts to seem more like a stage, an empty space for dramatic journeys, but also a regular living room to give the illusion of de-medicalising a medical treatment. A long way from the sterile white laboratories of clinical research, the living rooms are 'bourgeois dreams or simulations' of the everyday, as though to offset the weirdness that's on its way. The hoped-for effect of such tasteful domestication is *containment*: to house the unruliness of the psychedelic experience, to civilise it, to make it *useful*.

However, as Noorani also points out, 'the living room' has its own unruly history. In the nineteenth century the front room of the family home was known as the 'death room', where deceased family members received their final respects – as if to say, despite the best efforts of medicine to secure the treatment, the shadow of (mortal) risk cannot be avoided. It's a blind spot in the medical understanding of 'setting' I would only come to fully appreciate as my own journey unfolded over the coming months: that, however controlled the context for psychedelics, there was always a more or less obscure history (or histories) that might erupt at any moment. Really psychedelic settings were like piled-up strata in geology; medical science only paid attention to the surface layer, neglecting everything that lay below; earthquakes and volcanoes for sure, but also death masquerading as homely comfort.

Although Bede and I were prepared to tolerate a degree of risk, saw it as a necessary ingredient of change, I did not want the Dulces Sueños living room to be a death room. Our living room needed to *look* like a living room, but *be* safer: safe like a hospital. I removed some lethal-looking brassware between the couch and the rest-room. And if the welcome manual made it feel as though our home was under surveillance, so were the fabricated living rooms at Johns Hopkins. Noorani's description of the CCTV cameras and other monitoring equipment makes them sound like a Don DeLillo novel or an episode of *Black Mirror*. He notes the irony of celebrating psychedelic therapy for its 'breaking down of walls', while medical science dedicates itself to constructing and monitoring secure units, as though to communicate to a broader public that the drugs and drug experiences are not going to 'spill out onto the streets'. My concern was the opposite. I emailed our hosts that we had serious 'pampering' in mind this weekend and should not be disturbed for any reason. I lowered the blinds. I needn't have bothered: the entire community was nowhere to be seen; either Pine Mountain Lake's 2,769 residents were weekending away or they were conducting their own in-house experiments.

At the time of Noorani's research the living rooms had an altar of sorts, prettified with cut roses, scented candles and a figure of the Buddha, and the experiences were thought of as 'fruits or offerings'. In recent years explicit religious iconography had been removed so as not to 'bias' the setting. But my patient was already biased; he was a Christian monk. I instructed Bede to build a small altar on the coffee table: a simple wooden crucifix, a candle and an icon of his favourite saint and namesake, the Venerable Bede. At his request, I would sit in an armchair unseen at the head of the couch, psychoanalysis-style, with a notebook and a remote control for the music system.

Despite the fact that humans had been eating psychedelic mush-rooms for many centuries, the 'living rooms' at Johns Hopkins had a paternalistic flavour, as if to say: only safe evidence-based practice is legitimate. As with ketamine, there was an economic component to this. One consequence of the living-room trial design, notes Noorani, is to provide a rationale for 'commodifying not the drugs per se, but whole therapeutic protocols'.[3] The 'living room' is branding and IP for a whole *theatre* of psilocybin intervention that could be sold and rolled out into healthcare systems.

The overlap of business interests and psilocybin dates back to the Fifties, when Gordon Wasson, then Vice-President for Public Relations at J. P. Morgan, brought mushrooms to America following his experience with Maria Sabina and the ancient Mazatec tradition. ('The story of a wealthy North American investor going on a trip to the jungle to drink medicine, then returning home as a psychedelic evangelist is not as contemporary a phenomenon as we might think!' as Josh Hardman puts it.) Seventy years later it's not uncommon for Wall Street alumni to be running biotechs or managing funds investing in R&D for psychedelics-related enterprises.

Many companies currently compete for the intellectual property rights to templates of psilocybin-assisted therapy for the treatment of depression and several other psychiatric diagnoses. To make them patentable the chemical structure of the mushrooms themselves may need to be altered. Companies like Mindset Pharma are able to artificially synthesise psilocybin efficiently and patent the procedure. In the field of full 'mushroom therapy' COMPASS Pathways is the largest and furthest along. Established in 2015 as a charity, it flipped into a private for-profit some time later and, as the journalist Olivia Goldhill has reported, much of the data and intellectual property ended up owned by the for-profit entity, leaving many of the researchers who had devoted time and expertise to the development of them without remuneration aggrieved to see their work commercialised. In recent years such flipping from NGO to private concerns has become increasingly common in the psychedelics sector. But it's not all gone COMPASS Pathways' way. When the company released the top-line results of its trial of psilocybin on patients with treatment-resistant depression in November 2021, the stock plunged almost 30 per cent, reportedly prompted by the somewhat-middling results of the research, but also the relatively high incidence of serious adverse events.

The 'mystical experience' is the holy grail of Johns Hopkins' psilocybin-assisted treatment of clinical depression. At Imperial the equivalent experience is termed 'ego death', reflecting the Freudian orientation of the lead researchers. You might say that Hopkins owes its version of the concept to that other presiding god-father of our minds, the psychologist and student of religion, William James. As far as Michael Pollan is concerned, they are functionally equivalent. (Interestingly enough, both men were narcotic self-experimenters, Freud with cocaine, James with nitrous oxide.) For Johns Hopkins the mystical experience is the most 'active' of psilocybin's 'active ingredients', the implicit endpoint of all the medical engineering, the 'living room' theatre's great *reveal*.

In the 2018 study of high-dose psilocybin versus SSRIs in the treatment of major depression, 60 per cent of participants had mystical experiences, including Rachael Peterson. Many of the participants described it as being one of the five most significant events in their lives. Researchers at Hopkins defined 'mystical experience' using criteria from the 1960s established by the psychologist William Stace. These included qualities such as 'ineffability', 'transcendence of time and space', 'paradoxicality' and 'a noetic quality' – meaning a sense of revelation, as well as a deeply felt positive mood.

In foregrounding mystical experience, Johns Hopkins was reprising a finding from the famous 'Good Friday experiment' of 1962 conducted by Walter N. Pahnke on Good Friday at Boston University's Marsh Chapel. Pahnke, a graduate student in theology at Harvard Divinity School, designed the experiment under the supervision of Timothy Leary, Richard Alpert and the Harvard Psilocybin Project, to investigate whether psilocybin would act as a reliable entheogen – a synonym for psychedelic, literally 'god-embodying' – in religiously predisposed subjects. Nearly sixty years later, Johns Hopkins had secularised the church to a living room, but the same

Christian framework remained – hidden behind the sofa as it were, or covered by the fireplace rug, unobscurable, as though stitched into the DNA of the 'mystical experience'.

No need to secularise when the patient was a Christian monk whose devotion was woven into the fabric of his daily life: the monastic hours requiring him to gather with his brothers several times a day for the recitation of psalms, Eucharist, sacred reading, meditation. It was expressed in his work: giving spiritual direction to the clergy and those of other religions in his psychotherapy practice, teaching courses in clinical psychology and monastic spirituality at the Graduate Theological Union. And it threaded through the spaces in between these things, in the compassion, charity and *communitas* he brought to every encounter. His depression, strong as it was, remained for the most part masked, a pernicious, life-sapping secret. Equally his faith, beyond his daily observances and devotions, remained a riddle to me, and I know him well: twenty-five years ago he was my novice master for the two years of my brief monastic career, and we've remained close ever since.

Father Bede was thirty when he entered monastic life. At the time he was an intern at the Menninger clinic, treating patients with multiple personality disorder while he completed his clinical psychology and psychoanalytic training. Now approaching seventy, he had enjoyed a full and rich life. He had worked on severe mental illness, palliative care, the development of suicide prevention protocols for the US Military and as a consultant to monasteries throughout the country. His monastic career had begun at a Midwestern monastery before transferring to an Italian order of hermits located in the Santa Lucia Mountains of Big Sur. In the last few years he'd been made the prior of a small urban monastery in Berkeley of the same religious congregation, the Camaldolese. He was a talented clinician, a dedicated monk: affable, playful, smart and wise. In his spare time he was a keen amateur painter, a lover of Celtic folk music, a devotee of Korean food.

His depression had come on in his late thirties, and in an insidious, brutal way has made him its home ever since, as if the proximity to so much suffering in his patients had been contagious. The medication regimen had begun in earnest twenty-five years ago with Prozac. Over the years new psychiatrists had added in different SSRIs, amphetamines, anti-psychotics and cognitive enhancers, according to preference and diagnostic refinements. He'd stopped the Adderall two weeks ago in preparation for today, but he remains on two types of SSRI. Frank about the fact of his depression, Bede was less open about his internal states. Years of psychoanalytic training had made him a master of deflection, never unguarded, adept at meeting the other person on their terms rather than his own. Only in the briefest, un-self-conscious moments have I ever seen him sad. To my knowledge Bede had never had the kind of personal theology that would have God protect him from illness, or allowed himself to think that life has no purpose. He dealt with his symptoms – the reduced 'affect' (medical-speak for the expression of emotions), tearfulness, hypersomnia and insomnia, anhedonia (flat mood, lack of joy), over-eating, cognitive slowness, lack of motivation – with the same forbearance and uncertainty as the unbeliever. In fact, it might be closer to unbelief than that, because I suspect that Bede's faith was depressed too, buried under the freight of monastic life, though he wouldn't admit that, least of all to himself.

His clinical history coincides with the decades of stagnation and confusion in psychiatry; one might diagnose the profession itself with treatment-resistant depression. Until, that is, it was discovered once again that a few grams of psilocybin given in living-room-like conditions with a well-designed playlist could be an effective treatment. To account for its apparent effectiveness, psychedelic researchers have begun to develop their own theories of mental illness. For example, in March 2023, Robin Carhart-Harris and his colleagues published a paper[4] proposing a single factor as the putative mechanism underpinning various psychiatric disorders, including major depression. They call it 'canalisation', a term taken from genetics, which means something like 'over-learning', referring to a limiting of one's behavioural repertoire – getting stuck in a rut (or canal), one might say – as a defence mechanism against uncertainty. It involves compromising the brain's 'change capacity', its plasticity.

A key concept in psychotherapy is 'transference': the process by which a patient projects their own feelings onto the therapist. This is unavoidable, indeed important – by locating one's feelings of inadequacy, self-loathing, doubt outside of oneself and in another, the patient may become better able to grapple with and confront them. The flip side of this is counter-transference: when the therapist projects their feelings or desires onto the patient. I mention this because I clearly want Bede to have a mystical experience, to dig himself out of his canal, to reboot his faith, resurrect his Christian life. That's why, after weighing it carefully, I agreed to be his mushroom therapist. In clinical terms I'd be 'over-invested', making it important to watch the 'counter-transference', to keep my wishes to myself. Then there were the counter-currents of Bede's transference: on the one hand his keenness to try psilocybin, and on the other, the feeling that his expectations for today (for any day) were low. Both might be problematic.

As the last chapter's events had taught me, raising his expectations was likely to enhance the effectiveness of the treatment. I also knew that Bede trusted me; apart from our time together as monks, we speak regularly on professional matters, and we've even worked together, co-hosting retreats on meditation and neuroscience. The broader psychotherapy literature indicates that a good therapeutic alliance is a key active ingredient in treatment. We'd had our preparatory session: discussed the therapy's rationale, the light Christian framework, the music, the dose, the aftermath. Our setting was a comfortable and safe one (notwithstanding the distance to the Emergency Room). All told, the main factors relevant to psychedelic medicine had been covered, the placebo maximised as much as possible.

In addition to reviewing the scientific literature, I'd researched the guidelines on delivering psychedelic-assisted therapy. Oregon, a couple of hundred miles to the north, had passed a measure by ballot in 2020 that legalised guided psilocybin sessions in the state (they remain illegal federally), and would begin issuing licences in 2023. Given Oregon's reputation as a global mushroom hub (thanks in large part to its resident pioneering mushroom evangelist Paul Stamets) and all-round cult-friendly state (thanks to Sri Rajneesh), I was not surprised when I pulled up the government website for Psilocybin Facilitator Training Programs to see the listings for approved centres had names like 'InnerTrek', 'the Synaptic Training Institute' and the *'renowned'* 'Mending Mindcelium Method', whose website has a home page decorated with different kinds of giant, deep-brown shrubs, probably inedible, possibly poisonous, and boasts of having remained true to the 'original vision' – since foundation in 2022. I clicked on the one that sounded most serious, academic and reputable – the only one, in fact: the Berkeley Center for the Science of Psychedelics. For $10,000, it offers a programme of over 175 hours, 150 of them instructed, including weekend immersions, small groups, online learning, contemplative practice and a five-day retreat. I gleaned what I could about good practice from such websites.

But the decision to license therapy in Oregon was not proving straightforward. The decision having been made by referendum, constitutional challenges were emerging. One training organisation recently pulled out citing concerns about liabilities for therapists, who may be putting their licences or insurance at risk by participating. There was even talk about the possibility of the mushrooms being recriminalised in the months to come.

Bede took a little over 5 grams of *psilocybe cubensis*, four crumpled mushrooms dipped in manuka honey to sweeten the taste, at 9.05.

Earlier in the morning we'd discussed his 'intention'. Given the association of psychedelics and heightened suggestibility, presumably linked with neuroplasticity in the brain, great stock is generally placed on intention-setting, even without direct evidence of its effectiveness. Here the need was 'psychological': as a handrail through the abyss, or thread in the minotaur's labyrinth, or as a kind of magic voluntarism, that dreams may come true; or, more honestly, as a password for 'submission', in the knowledge that something uncontrollable, beyond all agency, was on its way.

Frank, the lay monk and psychology professor, was hounding Bede about it. 'Make it terse.'

'OK.'

'Make it specific.'

'OK.'

'Make it simple but hopeful.' The ecstasy of the true believer burned in Frank's eyes.

'OK.'

'But realistic . . . and not too specific.'

I interrupted. 'Can you go to your room, Frank?'

'Terse. Simple. Realistic. Specific. Hopeful . . . Oh, and put the *intense* in *intentional* – really go for it.'

After Frank had left us in peace Bede came up with the following: 'I would like a new way of experiencing myself and the world that's not so limited . . . My depression has become the background for all I see and feel, invisible to others, invisible to me . . . I'd like a different background. Something new. Or re-discovering what I already know, which might be one and the same thing.'

It wasn't terse, but it was moving. It also felt too cerebral, too rehearsed, as though for my benefit; part of our 'living-room' theatre, in other words.

9.32. Half an hour after taking the psilocybin. Bede is lying on the couch wearing an eye mask, draped in a white cotton sheet, still as Jesus in the tomb, listening to the sound of Tibetan bowls, while I sit a few feet away making these notes. I am dressed more formally than normal, as I would for work, with a notepad, a timepiece, recording observations – props of containment – to lend the veneer of science to the therapy.

I think about death and the living room. When the monks of Bede's order die they are placed in open caskets, made by the brothers from local redwood, wearing their white cowls. For twenty-four hours the open caskets rest in a chapel where live music is played – simple plaintive guitar chords – while the brothers take it in turn to keep watch. There's no embalming, so the observer sees the dead monk's cheeks continue to sink, the skin sallow, nails blacken. If the body is moved, as is sometimes necessary, and the sacral nerve accidentally stimulated, the dead man will involuntarily, holily, cross his arms across his heart. 'Men are afraid of death, the way children are afraid of the dark,' says the masterful historian of death Philippe Ariès.[5] The monks are at least courageous enough to look at it with the lights on. After the vigil is complete, the dead monk is buried directly into the mountainside without the casket.

The playlist has moved to India, a woman singing in Sanskrit over a droning harmonium. India: a country that understands death perhaps more deeply than any other. For two years I volunteered at a paediatric hospital in Varanasi, a stone's throw from the *ghats* where hundreds of corpses were burned each day. The whole of monastic life is, according to the rule of St Benedict, a preparation for death. I've since met dedicated psychonauts who say the same things about their practice.

9.47. Forty-two minutes in: he should be coming up by now, though there are individual differences in the rates of metabolising.

9.55. Against my explicit request Frank comes in. He looks worried. He summons me over and whispers his concern: he's found a thread on Reddit that suggests that for some very depressed people psychedelic immersion can be so strong that they can make an active decision not to come back, to let go of the old life, to let go of all life. I thank Frank and send him back to his room. Bede is smiling, the smile of someone choosing death. For a moment I wonder if I'm participating in the first known psychedelic-assisted suicide, and I imagine what several decades of life might be like in an American federal prison. I gently squeeze Bede's arm. 'Hey . . .'

He lifts the mask. 'What's wrong?'

'I just want to say: however wonderful it is in there, promise me you won't stay for good.'

'Where?'

'*There* . . . Paradise. I need you to come back, OK?'

'. . . OK.'

'Thank you.'

Bede pulls the mask down. In no way am I reassured by this. By the same token, I am a little disappointed by how coherent my patient appeared, how un-*fucked* he seemed.

10.33. I realise quite how deeply attached I am to the hope that Bede will have a mystical experience. The more I consider it, the more I doubt Bede has ever had a mystical experience in his entire life. The structure of monastic life was there to promote intimacy with God, but it could also turn to bone, as it had with me. I've often wondered whether Bede's psychoanalytic training vied with the possibility of his faith; whether theoretical knowledge of ideal-isation or denial would dry the 'oceanic feeling' Freud associates with religious fervour, and turn faith's tide, leaving only the sound of 'its melancholy, long, withdrawing roar', as the poet Matthew Arnold so famously described it. Then I have the thought that, with all this Eastern music on the playlist, it's possible he'll emerge from his trip converted, a freshly minted sadhu, an Asri Rajneesh Pura, a pantheist. Even, heaven forbid, a humanist. Fuck! What have I done?

Bede had lived in a hermitage for many years with Father Bruno Barnhart, a visionary mystic. Bruno had grown up in the north-east, scientifically inclined and utterly uninterested in religion. That changed one day in a graduate chemistry class, when a test-tube exploded and permanently blinded him in his right eye. In that same moment he had a profound mystical experience, powerful enough to change the course of his life. A few months later he joined an austere Catholic hermitage in Big Sur, where he spent the next sixty years, until his death a few years ago. For Bruno monasticism was not a profession, but an archetype, a capacity that resides in every-one: in as much as we try to unify our lives around a centre, we all have something of the monk in us. It was a notion that appealed to anyone who saw themselves as 'seeking' something, including me. 'What if it's not a place that you seek, but every place?' Bruno would tell me. 'What if it surrounds you so that the problem is not of find-ing a way *to* it, but of finding the way *out of* the ways in which you are stuck? What if it is "the everywhere", and we are imprisoned from, blinded from, the burning reality that we reach toward every moment through the thick vertical bars of our mind . . . ?'

Bruno was more poet than scholar. To his mind monasticism was a species of resistance, against the imprisonment of certain ways of knowing, versions of reality that were really just illusions of control. It was an idea that had stuck with me throughout my career: we needed a fresh encounter with the divine or with our 'inmost' selves – an experience to apprehend and dismantle our stuckness – to be able to see what Bruno called 'the light of consciousness within us, in a clear space . . . We are surprised once again to be alive. The breath is fresh once again, alive in our nostrils . . . '6 His very words sounded 'psychedelic' – and this from a man who had been a one-eyed hermit since before Leary ever took his first acid tab.

10.45. Listening to this music without psychedelic assistance is hard work: New Age obfuscations, percussive whorls, ambient sub-limities, disenchanting chants, with titles like 'Cloak of Darkness', 'Still Point in Motion', 'The Garden of Mirrors', 'An Eagle in your Mind'. The playlist itself is the result of recent Johns Hopkins research, since in the first wave of psychedelic-assisted therapy music was seen as integral. In keeping with the then dominant psy-chodynamic (Freudian) models, it was a handmaiden to the drugs, helping with the undoing of ego defences and the promotion of insight. But, like the mystical experiences the music helped inspire, the playlists consisted almost exclusively of canonical Western clas-sics that flourished in a deeply Christian milieu: Bach, Vivaldi, Handel, requiems, glorias, masses, the composers either early apologists for the faith or, in later times, wrestling with the meaning of faith in the various stages of their post-Christian contexts – Beethoven, Elgar, Arvo Pärt.

These same playlists have been used in the clinical trials in psychedelic-assisted therapy during the current wave of research, seasoned with the odd New Age wave-breaker, and iconic crowd-pleasers like John Lennon's 'Imagine' and Louis Armstrong's 'What a Wonderful World' for the encore. Like the 'living-room' motif, it was part of the domestic fabric of the intervention. And like the constructed living room, there was something 'instrumental' or functionalist about the modern medical rationale for music. Recent imaging studies suggest a 'partial overlap' of music-processing centres in the brain and the areas activated by psychedelics. As with dreaming, music works on the serotonin receptors whose signalling triggers expressions of emotion, autobiographical imagery and thinking about oneself. Though the research has not formally acknowledged it as such (music is considered part of the setting, but it's the drug-action that is causative), music has clearly been a significant lever in the triggering of psychedelic mystical experiences. Given the playlists' explicitly religious nature and psychedelics' capacity to enhance suggestibility, one might expect these experiences to have a distinctly Christian flavour.

Recently, however, Johns Hopkins got dozens of psychedelic therapists to rate different pieces of world music, including more percussive, vibrational compositions, categorising them as 'pre-peak' or 'peak' in their intensity, and in this way built up a less monotheistic musical trajectory to accompany the trip. This was the playlist we were currently listening to. Despite its efforts at non-exclusivity and diversity, the music still felt try-hard and mechanistic, aiming to trigger something specific in the mind of the listener, to bring even the most mysterious aspects of experience under some form of containment or control, as though being 'sexy' was nothing more than repeatedly pressing an erogenous zone. It was programmatic in a way that ran counter to Father Bruno's sense of music as bound up with mystery: 'Our life is welling forth at every moment from dark depths, we are continually arising, flowing upward and outward . . . Usually our consciousness is peripheral, our knowing is a downstream knowing. But in music . . . we have what we need to find the hidden source.'

10.52. Bede springs up from his bed, Lazarus-like, and goes to pee. His movement is agile, his attitude blithe, not suggestive of the overwhelming turmoil and bliss, the once-in-a-lifetime revelation, I had been hoping for.

On his way back to bed he flashes me a smile, as though he wants to chat. Feeling cheated by my own expectations, I give him silence back. I've gone from being concerned about the proximity of an Emergency Room and the strength of my patient's death wish to feeling he's not taking this whole business seriously enough.

'I can see geometric patterns, the smell of lavender fields,' Bede says, as though trying to please *me* with his trip report. But there is a geometric print on the toilet wallpaper and the air freshener is scented with lavender.

I turn the music up, jumping to the 'Peak' section – to 'Whirling into the Light', 'Iyanu Surprises', 'Journey of the Whales' and 'Call of the Divine.' Hit him with the hard stuff, whatever it takes!

10.58. Bede is lightly whistling along, his foot tapping under the white shroud. I want him to be suffering – I want him to reach what Matthew Johnson, a Hopkins psychologist, has called 'that point of criticality, of letting it all explode – laughing like a madman, crying like a baby' . . .

11.15. Bede looks more a sleeping baby. The jig has stopped, his head fallen off to one side, his mouth catching flies. For all the precision engineering, 40 per cent of participants on the psilocybin trial for major depression did not have mystical experiences. But my patient is a Christian monk: these guys are supposed to have mystical experiences *for a living*.

11.22. Maybe it's the mushrooms that are the problem. Frank had procured them from a grow-your-own operation in Oakland and kept them in an airtight bag, but there was still a chance they had oxidised.

11.46. As Bach's transcendental Mass in B Minor glorifies our humble, well-appointed, godless living room (there were still Christian vestiges on the new playlist) – a moment I hoped would be the apogee of my patient's ineffable union – Bede peeks out from under his mask and says, 'I'm hungry.'

I turn the music off. Magic mushroom journeys are supposed to be four hours, with nausea and an absence of hunger characteristic. Bede's been going for a little over two and a half.

'That was pleasant . . . really quite powerful.' He sounds guilty under the weight of my expectation. 'And the music was out of this world.'

Bede, by contrast, is depressingly of this world.

During our brief 'integration session' afterwards he rated the intensity of the experience on a 4-point scale at '2', then changed it to 'between 1 and 2'. There was not much detail *to* integrate: a few muted colours, warm feelings, the occasional shudder, nothing more significant. He kept telling me how lovely the music was, how deeply relaxed and *contained* it made him feel.

It was a commonplace that psychedelics, given sufficient dosage, were all but guaranteed to give you an extraordinary perceptual experience. While Bede fixed himself pancakes, I checked the literature again. Strong claims were made for the efficacy of psilocybin in treating 'intractable' or 'treatment-resistant' depression, but as we have seen, the criteria had a surprisingly low threshold: having used any two different antidepressant meds, even if they were from the same class, qualified for one criterion, 'polytherapy', and a minimum duration of depression of six weeks qualified for another, 'chronic'. Bede had been on an unstinting script of multiple different classes of psychotropic treatments for more than twenty-five years, as 'expert' in his field as any hardened Sixties-era psychonaut. And there were millions of others like him. The neurochemistry, the tolerance and sensitivity of Bede's brain were likely be of a very different order to those of patients at the low end of the diagnostic threshold. But to date, these variables had not been investigated by the literature. At the margins there were shreds of evidence pointing towards the possibility that neuroatypical subjects might have different responses to psychedelic drugs, or that the sensitivity and tolerance of those with a prolonged history of trauma might have been affected. But for the most part the narrative of the clinical literature was consistent and generalisable: the drugs all but guaranteed a psychedelic experience, and for clinically depressed patients this experience was most likely be therapeutic.

As the treating clinician I considered other factors that might have contributed to the underwhelming mildness of my patient's experience. Bede's psychological profile was complex and unusual. Years of personal psychoanalytic training had seen him develop refined ways of personal containment, enabling him to recognise how his personal emotions in response to the acute trauma of his patients might inflect his understanding of that trauma. He'd spent decades withstanding the powerful effects of other people's moods: might he therefore be able to *defend* against the powerful effect of the mushrooms in a similar way? Then there was the significant contribution of the placebo effect in psychedelic-assisted therapy. Might it be that Bede's depression, and decades of treatment failure, had created negative expectations (conscious or otherwise) that were strong enough to counteract the pharmacological effects? Both hypotheses, while plausible, were highly speculative and distinctly unscientific. Still the drugs hadn't worked, so my suspicion turned to the mushrooms themselves.

12.32. My clinical responsibilities over, it is time to switch hats, from facilitator to participant. With Bede safely snacking in front of *Succession*, Frank and I drive round Pine Mountain Lake looking for a trail head that isn't obvious from the map. Twelve minutes earlier, we each ate one broad-cap *cubensis*, weighing approximately 1.5 grams – less than a third of the dosage I had given Bede. This was the experiment's control: either Bede was insensitive, or the mushrooms were duds. The plan was to take a nature walk, a medically un-investigated form of psilocybin-assisted therapy that smashed through the walls of the living-room. Given the rate of metabolisation (approximately forty-five minutes for magic mushrooms) we should have ample time to find our spot, ditch the car, and surround ourselves with native forest in time to come up.

Unlike Bede, Frank was not clinically depressed. Rather he was professionally miserable: under the tough carapace of a grim Cleveland Catholic blue-collar pessimism, there was a softer, grimmer, more universal pessimism.

12.38. I continue to drive us round the rat's maze called Pine Mountain Lake, unable to find an exit. Frank is tetchy, grousing about land-planning in gated retirement communities: How could *anyone* find their way around this place, never mind the old and dementing? He's wishing he'd stayed indoors with Aquinas's *Summa Contra Gentiles*, which he was really *enjoying*.

12.41. We turn onto a highway and see a sign saying 'Yosemite 12 miles'. That's fifteen minutes. If the mushrooms work, we should still be just about straight enough to park the car, freeing us to enjoy some of the most spectacular scenery in the Western world. To a Brit, and a former literature graduate, who still believed in properness when it came to the English language, this was one of the few remaining things that really was *awesome* in this country.

'Just imagine what sort of poetry you'd have, Frank, if Wordsworth and Coleridge had been born in San Francisco and holidayed here . . . Just imagine what sort of nation you'd have.' I can feel the distant tingle of mushrooms in my unusually direct tribalism.

Frank looks unimpressed. In fact, he looks *awful*. 'We should get a move on.'

'Hit me with some America's finest driving music.'

Frank puts on some Gregorian chants.

12.52. There's no denying it: these mushrooms really fucking work! And much quicker than it says on the 'label'. I feel the tremor of an old delinquency I last experienced more than twenty-five years before, when I was drinking and drug-taking on a semi-professional basis. I also feel as though I am being forced back and down into the car seat by extra gravity, as though I'd taken a wrong turn and we are driving towards the earth's core. There are swathes of denuded fir trees on the hillsides, their trunks blackened at the base, a neat death in perfect rows as though someone has planted giant upside-down burnt matchsticks: Yosemite 2041, after the world has incinerated, to the score of apocalyptic Middle Age chants. It's like being on the set of *The Northman*. Or rather, *The Northman* has become part of my set, more chewing gum I couldn't get off my hands.

I tell Frank he's going to have to drive. He looks as though I've just told him he's going to have his prostate removed. Again.

13.04. Frank is a changed man. His eyes are filled with tears *and* he can't stop smiling. A pious, over-educated man, he's swearing like a Cleveland steelworker as we arrive at the National Park's admission gate. 'Damn-fucking-right-I-am, sweetheart,' is his response when the young ticket clerk asks him if he's old enough to warrant a concession. Frank is flattered but misguided. He looks eighty, is sixty-nine, and concessions start at sixty-two. Later that day I will be asked for proof I'm twenty-one when I try and buy kombucha at the local store. Psychedelic age, as slippy as snowy roads, which have meant it's taken nearly half an hour to get this far.

I need to get out of the car desperately, existentially.

I ask the clerk where we should park. She tells us it's another sixteen miles of winding road to climb up and then down to the car park on the valley floor. I think this is terrible news, that we won't make it.

Frank goes the other way: 'That's just fucking great! Tell me, which lane is the magic mushroom sightseeing tour? And put the fucking lane bumpers up, will you, darling? We're a couple of lame-ass beginners.' Driving has turned into human ten-pin bowling as Frank speeds off, laughing maniacally.

13.12. The mushrooms come in waves, like sharks. There are three feet of snow banked up on the roadside; behind it the conifers look illegally green. Frank and I are clinically under-dressed; we will die if we walk in this. Still, I have the overwhelmingly powerful need to get out of the car. Such escape behaviour is an ongoing feature of my response to psychedelic containment; any container is also the wrong container – too small, too big, too strict, too informal, too peopled, too empty. It always carries the charge of somebody else, an idea that's not mine. What I want is a leaky container, or a container where I am the leak. The road straightens out into a French-style avenue which seems to curl up into the white skies.

'Why is no other bastard exploiting the near limitless freedom the road affords?' asks Frank, slaloming between the white centre lines, then shifting the car abruptly into the left lane. 'It's just so much fucking nicer on the left, like we're in the "howling Highlands" of Scotland or the "finest-quality cotton" of the Indian Himalayas.' His accents are both acoustically and politically incorrect, though he is accurate about the leftist driving laws in both countries. The famous home-made psychedelic pharmacist Sasha Shulgin often used the 'drive test' to measure the strength of his new synthetic creations. I can now add my own personal observations: a car speeding on the wrong side of icy roads is not the safest milieu for psilocybin-assisted therapy.

13.25. We turn a corner to see a mile-high escarpment of limestone; every facet of the rock, every one of a million angles and flatnesses, 'pops out' like a colossal stone Transformer. It's so real and looming I duck.

'Feel how good the fucking air feels running through your hair, man.'

'The windows are shut. You have no hair.'

'We are pio-fucking-neers! . . . Mark my words: everyone will be driving this way before long.'

I have the thought the *car* is a Transformer, that I am its hostage.

13.45. We park in the valley. Miserable Frank is beaming, waving at children, talking to squirrels and birds, a reincarnation of St Francis. He'd never done psychedelics until a year ago. This was his third trip in the last three weeks. The literature suggested that the effect of a single large dose is long-term, if not permanent. Advocates contrast this with SSRIs and other conventional psychiatric drugs, which require daily use. But in Frank's mind he returned to his baseline misery after a few days and needed regular shroom-boosting. Even though legal usage seemed likely to become widespread imminently, I imagine many people would, like Frank, prefer their therapy non-medical, non-assisted, out of the living room, especially if it meant they didn't have to submit themselves to arduous, expensive, pathologising psychiatric assessments which might re-weight their health insurance and frame all future encounters with physicians.

13.55. 'You see that Netflix film about the guy who climbed up there without a rope?' We are standing at the foot of El Capitan, craning our necks.

'Don't be so fucking stupid.'

'It's true.'

'Nobody does that . . .'

'Alex Honnold did.'

'He *must* have been high.'

'*Free Solo*.'

'Is that what that is? . . . I thought it was a stupid *Star Wars* spin-off . . .'

Magic mushrooms. Pious, uptight, unhappy Frank turned into George Carlin. He began laughing hysterically at the thought of himself trying to climb up a tall lady he knew called *Dawn Wall*, and I caught his giggles . . . That was another thing that was missing from all the literature; the elephant in Johns Hopkins's sombre, responsible living room: just how stupidly, uncontainably funny these things could be. Punkishly funny. *Anarchically* funny.

Carhart-Harris and Karl Friston's 'anarchic brain' model – also known as relaxed beliefs under psychedelics, or REBUS – is based on the observation in brain scans of the 'entropic' or spontaneous freeing of activity in the cortex, the area associated with everything from consciousness and emotions to thinking and memory.[7] From this they derive a principle of how the drug works: psychedelics relax the precision of beliefs ('high-level priors'), thereby liberating so-called 'bottom-up information flow', particularly information associated with the limbic system, itself associated with more primal activities such as the fight-or-flight response, reproduction and nurturing. In other words, the brain on psychedelics is permitted to range freely and spontaneously over different states in an unpredictable way. It breaks the banks of the depressed, 'canalising' brain, relaxing the unhelpful priors which underpin various expressions of mental illness, flooding it with new possibilities. But only, according to the authors, if the context remains guided, with the 'right intention and care provision'. There is anarchy then, but it needs to be of the safe and thoughtful kind, 'medically supervised' anarchy . . . Hardly 'punk'.

'We're all Alex Honnold, Frank,' I said, emboldened boulders and furry firs dancing before my eyes, 'without a rope, nothing to hold, every moment a high-wire act.'

'Which is it? Rope or wire?' said a satanic-looking gnome in a yellow beanie: the real Frank, Frank's *geist*. 'When it comes to counting, it's chickens, not blessings. There's no such thing as a *free* solo.' A gnomic gnome.

In retrospect, there were three things that gave us away as different from the cartoonishly bright Patagonia Teletubby tourists that teemed on the valley floor. First, we were the only people not taking photos. Instead, we watched in a state of befuddlement the peculiar modern geometries of a woman taking a photo of a daughter photographing a deer that had lost its iPhone shyness so completely it was licking the girl's camera, giving her a 'tongue close-up'. Second – noticeable to the more stationary tourists – we spent more than two hours walking the same fifteen-minute loop and yet, like brain-ablated mice in a simple maze, or older residents at Pine Mountain Lake, we did not learn or remember anything, viewing each lap with the same astonished incredulity, the same innocent wonder, as though the secret of life was to want what you already have, again and again and again. And lastly, we were the only people yawning – jaw-dislocating, mushroom-inspired yawns that folded the whole body in half – before the most spectacular scenery on the continent. Every so often one of us would look up and the nonsense conversation would fall silent; bodies turned inside out at the sight of Half Dome or Yosemite Falls; the cliff faces moving through us or 'discovered by the right body set', as Merleau-Ponty might have it, searching, like Alex Honnold, for 'maximal grip'.

Nature has always been a regular setting for psychonauts. If their psychedelic usage was ever 'therapeutic', then they preferred their settings to be 'ecologically valid'; authentic real-world locations, that is, rather than synthetic set-pieces like Johns Hopkins' living rooms. ('Personally, I like to live in *all* my rooms,' I could hear the ghost of legendary psychonaut Terence McKenna saying.) So far psychedelic research had paid only lip service to nature and ecological validity. Though the therapy itself has always formally been indoors, participants have been found, via questionnaires, to increase their sense of 'nature-relatedness' either side of treatment. 'From Egoism to Ecoism . . . ?' was the title of one of Carhart-Harris's Imperial papers. The researchers want to imply a causal order: psychedelics change the set of the consumer, predisposing her to certain types of settings: green ones. Maybe. But then the two might just go together well; you wouldn't say salt causes vinegar any more than cheese causes onion. And if the relationship is causal, why has there been no attempt to incorporate nature into settings for therapy, especially when the nature is as sublime as this?

Of course, 'nature-relatedness' is no less a construct than the living rooms of medical therapy or the 'mystical experiences' with which it is associated. There are so many different ways of conceptualising nature across history and different cultures, from man's bucolic playground to his violent negation of it, so many different ways of *relating*, but in psychedelic science nature-relatedness is all of a piece, another technology for improving our mental health.

At this point, I – anarchically – want to propose that the medical model for explaining the 'profundities' of psychedelic experience is a bowl of cold, half-cooked, semi-digestible mushroom stew, into which have been thrown, at different times, 'ego death', 'mystical experience', 'nature-relatedness', the Romantic Sublime (though not psychedelically 'informed'; narcotics were often involved) and awe. Awe – cornerstone of the Great American *Awesome* – has had something of a reprisal in recent years thanks to social psychologists like Dacher Keltner and Jonathan Haidt, who have located it 'at the upper reaches of pleasure and on the boundary of fear'.[8] It's viewed as having the potential 'to significantly alter the self-concept, in ways that reflect a shift in attention toward larger entities . . . and the diminishment of the individual self'. Which sounds remarkably similar to the model for understanding psychedelic therapy or mystical experience or the sublime. This soup's flavours are indistinct, its ingredients logically overlapping: psychedelic states may create awe, either in nature or in an artificial living room, but then nature may create awe in the absence of psychedelics. Equally, all of them might be good for our mental health and provide the setting, either alone or in conjunction with one another, for a mystical experience which would be *really* good for our mental health (unless it shifts our underlying beliefs away from what's rational: then it might be dangerous). Soup, I say.

Because of these overlaps, the logic of what psychedelics actually do, and indeed what the nature of reality is, becomes murky: are the drugs some kind of hack on poor attention, catalysing our ability to see what's always there? Or is the psychedelic brain generating the awe, lending emotional freight to what is otherwise mundane? Or does the quality of awe belong in some way to the objects themselves, to Half Dome, or Mont Blanc if you're Shelley, or to red dwarves through the Hubble telescope, or the spectacle of millions of baptising Hindus at the Kumble Mela? Or does the awe rest, in fact, in our descriptions of these experiences, either because we find ourselves defeated in the attempt ('How was it?' 'Er, I dunno . . . I mean . . . It was just . . . awesome'), or because we triumph (or are triumphantly defeated), such as in the poetry of Shelley and his compadres.

And if awe does belong to the objects themselves, as the phenomenologists might say – in their steepling grandeur, their incontrovertible majesty – then why is it that I'm feeling awe now, transfixed by the small wastepaper bin at the foot of Yosemite waterfalls; the sheer complexity of the ironwork, the heartbreaking detail of its patterned rust, which to my mind looks as inexpressibly complex and compelling as the Milky Way, which I rarely notice? The bin is no way overwhelming in its power or scale. It might be in nature, but it's certainly not of nature, it's frankly inorganic, except for the fact that it's abundantly alive to me now; fragile, complex, moving. I never want to leave its presence. If this moment shared somehow between bin and me is on the 'boundary of fear', it is a fear that such enthralling objects are available everywhere, at all times, that anything and everything is equally mysterious, that all reality is always being psychedelic: it's just that the brain's salience network, unmoved by mushrooms, has deemed reality unworthy of its attention . . . Meanwhile, I can't stop weeping; the thought that, after those first wonder-filled years of childhood, I had spent the rest my life failing to really see things, failing to love bins. What waste; my life its own waste basket . . . That small, nondescript trash can at the foot of the stupendous waterfall brought it all home.

16.30. Frank and I talk about Bede; our love and admiration. As true as it is, talk of Bede is also the funnel through which both of us can open our hearts fully, show the extent of new-felt capacities of feeling and expression, unsayable under normal circumstances, especially between two well-defended, ageing men.

16.45. Frank is coming down, getting cold and hungry. It is time to drive home. Meanwhile I am still in it up to my eyeballs: arriving at the car, I have to overcome the thought that the vehicle might suddenly reconfigure itself as Optimus Prime.

17.20. While Frank eats non-conceptual soup by the fire I walk through the elm-bowered deserted lanes of the Pine Lake Mountain Estate. I listen to the same playlist I used with Bede – only now, in this set, with this setting. 'Journey of the Whales' and 'Novus Part 1: The Flying Bach' have, like the litter bin, become improbably moving. (In some remote nook of the living room of my mind the spectre of drug-induced bad taste was sleeping.) I don't want to lose this newfound sensitivity, which feels more real, more aligned with reality than before. Under the anger there is a terrible sadness: Bede's treatment hadn't worked, meaning the latest hope of relief from depression had gone the same way as all the other new medications he'd tried over the last thirty years, drug regimens that had blunted, taken away more than they had relieved. For Bede at least, psilocybin would join a long list of mothballed 'new' cures.

18.10. A lone car swerves off the road and crosses my path a few metres in front of me. It keeps going in slow motion, until the front wheels hang off the edge of the pavement, several feet above a steep bank of earth.

A young girl with a nose ring gets out and stares at the front wheels hovering three feet above ground. 'Shit, I wasn't concentrating.'

'Get back in,' I say. The plight of others has become so profound. 'Put her in reverse.' Sounding like John Wayne. I really must have thought I could lift the whole car. I push upwards on the front wheel axle with all my metaphysical belief.

Eventually a boy in a hoodie comes out of what must have been his grandma's house. 'What the fuck are you doing? . . . Are you stoned?'

I think he means me, but he is talking to the girl.

'Shit, we're gonna need a fucking tow truck,' he says.

Neither of them notice me clamber back up the bank and walk off. It was time to go home.

Bede had been the intended target of mushroom therapy, but if anyone had a mystical experience that day it was me. The awesomeness of Yosemite, the scintillating music, the openness in my personality, the relaxing of high-level priors in my brain, the sense of love in my heart (I had left poignant voice messages for more than a dozen people, some of whom I only knew from Groupchat, and extralong ones for Nigel and Lenny for giving me this wonderful *permission*), had all been abundant, just as the therapeutic literature predicted. But there was also anarchy, un-containment, breaking out of the living room, foul-mouthed delinquent humour, and most of all the awesomeness of that rusty fucking bin, which the literature hadn't predicted at all. Really it was all of a piece, pointing towards a revelatory sense of connectedness with others, within myself, with the world at large, that felt inspired. I'd go so far as to say that I experienced transcendence in a much more visceral way than I ever had during my years as a monk: for those few hours it felt as though my world view had been altered, shifted towards something more pagan than scientific.

It was all in accordance with the literature: such shifts are quite common in people who use psychedelics.[9] And the literature also told me that such non-scientific tinglings could be explained and understood by science, and that in fact nothing non-scientific had occurred or been experienced. The argument is made most clearly by the philosopher Chris Letheby, author of the recent *Philosophy of Psychedelics*,[10] who had collaborated with Johns Hopkins. The partnership was welcome, Letheby's materialist philosophy helping to make explicit and organise some of the underlying assumptions of the psychedelic research project. His book proposes that all psychological experiences can not only be located in the brain (what experiences can't?), but can also only be *fully understood* using the terms of brain science. According to Letheby, whatever I might believe about the metaphysical nature of my mystical experience, it could be explained in its entirety in terms of physicalism. The implication of all this was that whatever far-out views of the universe I now held thanks to my experience of unconstrained selfhood and newfound ability to construct the world in thrillingly new ways, the scientific world view remained intact.

This explanation seemed likely, but part of me wondered whether it also sold the idea of mystical experience a little short. I thought of Father Bruno, who would speak of the monk coming into an awareness that is open and flowing forth from within: 'Our life is welling forth every moment from dark depths, we are continually arising, flowing upward and outward . . .'

20.15. Later that evening Bede says a simple Mass for Frank and me at the living-room table. We are sombre, keen to downplay our time in Yosemite in front of our friend so as not to provoke his envy. In the months that followed Bede would say the mushrooms worked, just not in a pharmaceutical way. Their failure had jolted him into a new acceptance and freedom; he was no longer looking for miracle treatments. Meanwhile, the host and wine are in little cups, the rest of the mushrooms in a bowl nearby with the jar of manuka honey, one dream next to another.

3

Bad Trips

The ceremony was more than halfway through when Aurora picked up a drum and beat it under the *icaro*. The new rhythm brought people to their feet. Of all the possible ways to dance they were dancing exactly like one another, throwing the same bird-like shapes, eyes closed, before a pale gold Californian moon; there are levels to suggestibility. I remained in my seat, too sober, too aware of how odd this all looked. My neighbour nudged me: there was still time for a 'late sip'. Sometimes hell is other doses; whatever you've taken is too much, or not enough, and underneath that, the fantasy of perfection, what my new best friend Geo calls 'just enough of *too much*'. I could have ignored him and sailed peacefully across the rest of the night.

'A *pochito* of a *pochito*,' I requested, kneeling in front of the altar, pointing to Aurora's smallest cup. They say you can look into a shaman's eyes and see what's coming your way.

It was quick. Fifteen minutes later I knew I was in trouble. There were beautiful visions, but I wasn't *watching* them; rather they moved through my insides like deranged roots, like eels that might at any moment burst the walls of my stomach. *I wish I hadn't . . . I should have . . . I am such a . . .* Every time the silent 'I' surfaced in the speech of my mind, the eels got more agitated. 'You don't get to decide,' they whispered. 'This is our terrain, and the lesson here is death.' (While it's common for people to hallucinate indigenous animals, I'm from England: we don't really have serpents – eels are what we have.) I thought I saw ruefulness in Aurora's reptile-green eyes, and it turns out I was right. The drumbeat had picked up, everyone but me writhing on the dance floor like furious eels. Still I kept

trying to martial the chaos with half-formed thoughts. *If I can just—.*
I might be able to—. What about if I— . . . I— . . . I . . . I . . .

Finally I admitted defeat, wrote out a neat label in a clear hand
and stuck it on the jar of what was happening.

I'm having a <u>bad trip</u>. *That's what this is, it's a* <u>bad trip</u> . . .

And with that, my fate was sealed. For that night certainly, but
also as a floor effect, a terrifying prior that would haunt my difficult
moments in all future journeys.

For the underground psychedelic chemist Sasha Shulgin, madness begins with paying attention to the eels' whispers. 'Each person
has his own brand of toxic psychosis. Mine always starts with the
voices in my head talking to me, about all my worst fears, a jumble
of warnings and deep fears spinning faster.'[1]

While the brands appear infinite, my conversations with others
suggest the narrative archetypes are few: *Man believes he's dying, man*
swears he's losing his mind, usually fuelled by *man convinced his current*
state will persist forever, with combinants like *man would prefer to die*
than endure this and *man clings to belief he would be less frightened if he*
really did lose his mind, all of them constellating around the lethal
trap of selfhood. Convinced he is beyond help, man still maintains
that he is the only man who can save himself, as though self-
annihilating fantasies must contain self-hero-ifying ones.

In the Fifties psychedelics were often labelled 'psychomimetics'
because of their capacity to imitate psychotic states, or 'hallucinogens', emphasising their demonic potential, compared with the
more wholesome-sounding 'mind-manifesters' or 'God-embodiers'
('psychedelics' and 'entheogens'). In science as in any form of brand
management, naming is all.

For the next two hours I lost my mind with the fear I was losing
my mind. I tried the following forms of what clinicians call 'self-
care': asking for help, counting breath, replacing the thought 'Oh,
no' with 'Oh, yes', jamming my hand down my throat to induce
vomiting (I felt eel breath on my fingertips), replacing an atheist

curse ('Oh, my fucking God') with a Christian prayer ('Please fucking help me, God'), Buddhist meditation, running through the forest screaming, asking for more help, asking Aurora for the medicine that cures me of her 'medicine', performing body scans (these were like feeding time for the eels), asking a dog for help, 'playing' with different postures (on my knees, arms aloft ready for execution, the starfish, the spreadeagle, the child's pose, which regressed to the foetal position, which matured into the 'hair-on-fire' asana) . . . I panicked but I couldn't even concentrate on panicking. I remain grateful that nobody provided me with the gun I asked for, or the keys to their car, or the cell number of Michael Pollan.

It was the single most terrifying episode of my life. I got a glimpse into different expressions of insanity – somewhat like, I imagine, the acute psychosis of patients I have worked with. That was insightful, but too insightful. In fact, I felt as though I was experiencing the majority of the diagnoses in the *DSM* (*Diagnostic and Statistical Manual of Mental Disorders*; there are 300, more or less) on a repeat cycle, and though the cycles will have been farcically quick, the felt reality was, as William Blake says, of all our states, eternal.

And yet forty-eight hours earlier, in the same ceremony, with the same facilitator drinking the same medicine, I'd tasted only cosmic ice cream.

The Psychedelic Renaissance is built on the foundations of a revolution in mental health; on, that is, the untrammelled psychological benefits associated with half a dozen drugs. But the narrative is beginning to change. In November 2022, the *New England Journal of Medicine* published the results of a study conducted by the drug development company COMPASS Pathways, which it had originally revealed in November 2021, investigating the effects of its patent-protected 'COMP360' synthetic psilocybin on treatment-resistant depression. A hundred and seventy-nine of the 233 patients reported at least one adverse event (headaches, nausea, fatigue, or insomnia). Some consequences were considered tolerable 'side-effects', built into the price the clinically depressed patient might be willing to pay for relief. However, 7 per cent of the sample experienced so-called treatment-emergent serious adverse events (TESAS), including suicidal ideation and self-harm. Meanwhile, 12 per cent of the 11,000 participants in the recent Global Ayahuasca Survey needed professional help for coping with adverse experiences. A systematic review of adverse events in clinical treatments with psychedelics published in the *Journal of Psychopharmacology*[2] concluded that 'Adverse experiences are poorly defined in the context of psychedelic treatments and are probably under-reported in the literature.'

The focus of therapeutic research has been on different facets of 'positive' experience – 'connectedness', 'nature-relatedness', increases in creativity and openness, and their role in the broader concepts of ego death or mystical experience – rather than the fear and terror that may live, unacknowledged, alongside or within them. But trial participants, however 'light-filled' or 'abundance-manifesting' their 'intention', may not get to choose where to place their focus during their journey. Choice is not obvious when the thing you're dealing with feels as overwhelmingly powerful as internal eels, their only apparent purpose to cause you harm. Rachael Peterson, the Johns Hopkins participant who went from mystical experience to 'grief-tinged cosmic panic attack' over the course of two identical trials, has recently spoken out about the neglected shadow-side of treatment:

> The effort to medicalise psychedelics has focused on a narrow subset of experiences that are positive and therapeutic. Variations are dismissed as statistical outliers, flukes resulting from flaws in set and setting or vulnerabilities in the patient. A serious effort to examine bad trips can be perceived as positioning oneself 'against' the movement.[3]

Hype creates a particularly inimical position for dissident voices, where to speak out is not seen as challenging utopian hyperbole – its repression of the more indigestible elements of psychedelic experience – but as a damaging disservice to the 'movement'. It's the same kind of tribalism that's increasingly pervasive in the broader culture: if you're not on message, you're the enemy.

Despite the emergence of a narrative of 'medical harm', clinical research has put most of its efforts into looking the other way. The neuroscientist Matthew Baggott, who worked on a number of MDMA trials, has spoken about how research protocols either ignore data that points towards psychological challenges, or simplify it so that it best fits the existing model, making 'challenging experiences' an undifferentiated information bin. Implicitly or otherwise, the challenging experiences tend to be seen as 'side effects', partly because funding is generally contingent on certain types of research emphasis. The irony of this, as Baggott notes, is that these *side*-effects may, in some cases at least, be central to the sought-after effects of mystical experience or ego death.

Once again, what's lacking is an authentic, fine-grained curiosity about the experiences themselves. To call a trip challenging is, as Peterson notes, like saying running a marathon is 'hard':

> We lack the vocabulary to describe the vast diversity of experiences that might fall under the banner of challenging, difficult, bad, or adverse. Someone may describe a psychedelic experience as challenging . . . engaging in behaviour that could harm themselves and others during the session – swan-diving off a couch, for example. Another may use the same term to refer to intense emotions of grief, anguish, sadness or despair . . . It could refer to the frenetic encounter with weirdness, an experience that shatters preconceived notions of matter, agency, the cosmos. 'Challenging' may describe traumatic memories, confrontations with familial dynamics or painful insights. Someone may even have a 'meh' experience – uninteresting – but call this 'difficult' for not living up to some hoped-for breakthrough.[4]

Most people enter psychedelic space because they are motivated to disturb normal conscious experience in some way. Challenge is built into the rationale. Writing this after the completion of my own first-person research, I can say that I have never had a psychedelic experience that hasn't had at least a momentary difficulty associated with it. These difficulties may resolve themselves naturally, or they may intensify; some of them may turn, as Peterson suggests, into crucibles of lasting insight, or they may persist 'unproductively' long after the neurochemical journey. A small but significant number of cases may be left with a psychiatric diagnosis; for example, hallucination-persistent perception disorder (science for ongoing 'flashbacks') is included in the *Diagnostic and Statistical Manual 5* (*DSM*). Clearly this intensifies the already complicated ethical issues around consenting to an unknowable transformative experience. How should the patient evaluate a treatment that, while it might resolve long-term depression, might also change her metaphysical beliefs, cause geometric patterns to inhabit her normal waking life, and even lead to more intensive thoughts of killing herself in the aftermath?

Institutional airbrushing is less surprising if one thinks, as John Gray does, of the humanist philosophy that might be driving aspects of science: the simple belief that we can progressively improve or refine our nature with different forms of enhancement. The same underlying assumptions are driving the significant investment in psychoplastogens discussed in Chapter 1, designed to efface all experiential components, good and bad, as they are at institutions like Neuroscape. Interviewed on the Tim Ferriss podcast,[5] Adam Gazzaley, Neuroscape's Founder and Executive Director, described the development of trip suites in which participants are hooked up with multisensory feedback technologies (aural, olfactory, haptic) that may be triggered in response to changes in physiology or observable behaviour: the aim is to refine the trip, optimise it, make it more efficiently therapeutic by 'pulling micro-levers' on set and setting:

Her EEG looks a little stormy, turn up the dolphin song.

The cortisol's through the roof – ground him with Montana breeze and a puff of woodland musk.

The fantasy of opening up the black box contains a deeper fantasy still: taking the 'blackness' out.

Different religious traditions have long recognised the centrality of suffering to human experience, regarding it as a vehicle for spiritual development. The first noble truth of Buddhism's fourfold path is 'All is dukkha' (from the Pali for 'unsatisfactory'); think of Jesus's tears in the Garden of Gethsemane, or Job's meal of pain and pestilence. Medical science apart, there are underground psychedelic traditions in the West, and indigenous traditions in the South, which do the same (of which more later). According to this school, learning how to navigate adversity, how to cultivate tolerance, accept fear – recognise it as a trick of egotism, one of the myriad ways we deceive ourselves – is part of the psychonaut's development. This kind of psychedelic practice embraces the nuances of 'challenging experience', the different personalities of the demons and tricksters that one may encounter on the journey's way. (Countercultural devotee Erik Davis goes further and suggests it's a religion in its own right.)[6] It's the near contrary of that kind of science which depends on the containment of a passive patient, the chemical engineering of a benevolent event, the streamlining of suffering, the airbrushing of madness, the demonic and everything else that's undesirable. Unless, that is, what's undesirable is collectively characterised as 'trauma'; then it moves back to centre stage.

Trauma has become ubiquitous in the discourse on mental health, part of the living room's furniture. To paraphrase Will Self, one might think of it as a 'silent epidemic' if people didn't talk about it all the time.[7] It has become a bin for clinicians and laypeople alike, containing all that is intolerable, intractable, overwhelming, exceeding our capacity to reason. In the West we are in the habit of exporting it internationally, using it to interpret other cultures which have never encountered the concept: one paper estimated that 95 per cent of the inhabitants of Sierra Leone met criteria for post-traumatic stress disorder. We also export it historically in the humanities, using it to read our past, real or imagined: Hamlet's melancholy, Don Quixote's windmills, the life of nearly everyone in medieval England.

'Trauma, you think you've got trauma?' I imagine John Cleese's voice saying. 'I've dragged this sarsen stone all the way to Salisbury, carried it with my bleeding hands for two whole years over dale and moor and heath, and what have I got to show for it? Freedom? No chance. A spot in Christian heaven? Not on your life. An auspicious pagan sacrifice? You should be so lucky. *Trauma*, that's what I've got to show for it, that's what I'm going through according to those know-it-all cultural historians of the next millennium, the complete shattering of my world view, of everything I assumed was safe and reliable, leaving me susceptible to a horrible cocktail of dissociation and re-livings . . .'

'Well at least you've got trauma,' pipes up another peasant – it's Michael Palin. 'I lie awake at night in my frozen cesspit, dreaming of trauma . . .'

Such lives were objectively hard by modern standards, and undoubtedly distressing, but the impulse to pathologise, to make trauma an artefact of neurocognitive psychiatry, is our own, implicitly endorsing the same humanist philosophy that regards history as a progression towards reason and health. Accordingly, the trauma we have is less than what came before, and now we have the technology to overcome it. But progressivism does not equate to science itself: it's an (implicitly Christian) ideology and one that's specific to our time. 'In pre-Christian Europe it was taken for granted that the future would be like the past,' writes John Gray in *Straw Dogs*.[8] 'Knowledge and invention might advance, but ethics would remain much the same. History was a series of cycles, with no overall meaning.'

As per the third edition of the *Diagnostic and Statistical Manual*, published in 1980, trauma was officially crowned as its own mental health condition, post-traumatic stress disorder. The *DSM* is a clinical reference book, but it's also the enshrinement of a whole way of seeing: biological psychiatry's medicalising of specific behaviours and experiences. The following forty years and successive editions of the *DSM* have seen a huge increase in the prevalence of PTSD and a general diffusion of the concept. One can still meet criteria following discrete traumatic events – industrial, technological, militaristic, the modern equivalents of railway spine and shell-shock – but by the time of the publication of *DSM IV*, seeing others undergo them can also lead to a diagnosis. Or one can endure a numberless variety of smaller, adverse events – most likely in childhood – whose gravity can never be fully escaped, that are now bracketed under the diagnosis of 'complex trauma'. The expansion of criteria makes many of us, most of us, candidates. A person's 'trauma history' has become a major instrument in psychiatry's overall *set*, found to be highly correlated with depression, other anxiety disorders and addiction. Meanwhile our *setting*, contemporary Western society, with its various catastrophes and imagined threats, is increasingly thought of as *traumatogenic*.

Within the confines of psychiatry, trauma straddles, rather like psychedelics, discrete diagnoses ('transdiagnostic' is the technical term): one type of challenging experience is associated with different types of clinical presentation, which sets up the broad rationale that one molecule can treat them. (The tendency to equate one psychedelic compound with one diagnosis – psilocybin with depression, ibogaine with addiction, for example – may be a cultural artefact, a matter of research or commercial bias, as much as it is an authentic 'scientific' link, as we shall see in the next chapter.) The association between trauma and psychedelic therapy goes deeper still, to the conceptual level: psychedelic treatment has come to be understood as a kind of 'inverse PTSD': a single, overwhelmingly intense, emotional episode that has lasting *beneficial* effects on emotional processing.[9]

To date there has been little space in the literature for thinking of psychedelic experiences themselves as traumatic. My bad trip certainly echoed horrible, disintegrating feelings I'd occasionally experienced in childhood. It also overlapped with similar feelings I've had very recently with the different turns of my daughter's illness. The sense of powerlessness in the face of extreme adversity appears to go to the heart of trauma, taking us to the place where the humanistic belief in our own agency is compelled to break down.

For a while after that second night with Aurora I felt desolate and terrified, an experience which in some way cast its shadow over all future journeys. My limits had been exceeded; I felt powerless. In the Johns Hopkins trial many of the participants had described their mystical experiences as one of the most important of their lives. Well, this one came straight in, pound for pound, at number one in my personal Top Ten Darkest. What was also clear was that the challenge I had encountered was accompanied, perhaps compounded, perhaps even caused, by a desperate unwillingness to accept what was happening to me; an ego-death in reverse. I vowed that it would be the last time I ever took psychedelics.

And yet, unbelievably, *insanely*, you could say, two nights later I returned to the ceremony and took another dose, after two days of listening to those with more experience attempting to mentor me in my madness.

A bad trip may be the best trip.

You get the trip you need, not the trip you want.

Running away is the problem. Turn round, talk to the demon.

That third night I saw my traumatic experience within a larger unfolding narrative rather than as a single, terrible set-piece. From that perspective what had appeared to be dreadful, and mortal, one night, felt more like a cosmic joke the next – eels are strangely cute when you really look at them – in which, crucially, I don't get to decide the punchline. Though clichéd and somewhat contradictory, the psychedelic coaching I received from more experienced practitioners, Aurora in particular, was useful.

Rachael Peterson had thought differently, finding her New Age peers' positive prompts and optimistic re-framings of the terrifying experience paternalistic and unhelpful. 'I learned that to confess to enduring challenges from a psychedelic trip is to render oneself a dartboard pierced by a million speculated whys: latent psychological problems provoked by the experience, the wrong set and setting, the wrong medicine, the wrong dose, the wrong day, the wrong guides.'[10] Like any ideology, this shaping of experience via social incentives – to fit in, to be part of the gang, not to poop the party – is part of the group psychology that powers psychedelic usage.

'Tellingly,' Peterson continues, 'I have never encountered an analogous reaction to "positive" trips; when I share healing experiences, no one rushes to diagnose what enabled them to occur.'

I share Peterson's concern about this asymmetry. I also share her concern regarding the general, casual binning of challenging experiences, the assumption that they're some kind of humus for 'post-traumatic growth'. Even the concession that certain psychedelic experiences may qualify as 'traumatic' is medically circumscribed, as requiring further pharmaceutical treatment; what one pill gives, another is on hand to take away.

MDMA, an 'empathogen' rather than a classic psychedelic, is associated with heightened feelings of empathy, affiliation and interpersonal trust. It was not legal when I crossed over from California into Oregon for my third trip, MDMA-assisted therapy. The FDA had granted 'breakthrough therapy' designation for MDMA-assisted psychotherapy for PTSD, progressing to Phase 3 clinical trials in 2020. The initiative had been driven by MAPS, a non-profit organisation that had been campaigning since the 1980s. Two years later, Phase 3 is well underway, and initial trials have found that 67 per cent of PTSD sufferers who received treatment no longer met diagnostic criteria, compared with 32 per cent of those who received a placebo with psychotherapy. The clinical evidence was backed by data from brain-imaging studies: the drug dampens the activity of the amygdala, the brain region involved in fear, allowing people to more comfortably revisit and process terrifying memories. Studies on rodents, meanwhile, suggest MDMA can reopen a 'critical window', allowing the brain to regain the plasticity seen in younger brains that are capable of 'learning new behavioural responses'.

This was the latest incarnation of a drug I'd known about since the late Eighties. The last time I'd taken it was in 1991 at the Hacienda club in Manchester, in the form of Ecstasy, the drug of choice for rave culture, for making us fizz with 'oneness' to the sounds of the Orb and Spiritualised, a phenomenon sociologists would later call 'collective effervescence'. A few years on I would sit in graduate classes on neuropharmacology and hear of 'emerging data' that tied the same drug to 'negative changes in affect regulation' (apparently it made people miserable rather than ecstatic), and 'possible deficits in executive functioning' (it made people more stupid rather than cognitively enhanced). Twenty years later, little mention is made of that 'emerging data', probably because it was driven by a prohibitive culture and inferred on the basis of small sample sizes; an anti-hype train. Now the train goes in the other direction: while MDMA was known to be more toxic than the other psychedelics, therapeutically assisted doses for traumatised veterans in clinical settings had been gold-stamped by the FDA. After its wayward youth, it seemed as though the drug itself had been finally rehabilitated.

Then in March 2022, the same week I travelled from northern California to Oregon, *New York Magazine*'s second season of the *Power Trip* podcast dropped, and a new type of conversation began.[11] The season turned its attention 'above ground' to look at therapeutic abuses within MAPS – and their Phase 3 trials. There were bad trips and then there were *bad* trips; unsafe settings, predatory practitioners, the already vulnerable made more vulnerable. The same medicine that relieves traumas – that might one day heal the world, according to MAPS' founder Rick Doblin – was paving the way for the creation of new ones; a different kind of *gateway* drug.

If this was surprising, then it should be the surprise itself that was surprising. The history of psychotherapy is shadowed by abuses of power and therapist–patient relationships that turn dark when the therapist fails or even intentionally neglects to regulate the 'attachment' of patient to therapist that is fostered by therapy, and on which the process relies. Adding in a drug that made the patient more open, more desirous of connection, more suggestible, and therefore more vulnerable, would likely only enhance the risk.

A few weeks before I had spoken to a former colleague, Ruben, a recent convert to psychedelics, who had engaged a MAPS-trained duo – a female therapist and a male psychiatrist – to provide him with MDMA therapy for complex trauma. Ruben was now in his early fifties, but profoundly difficult early experiences continued to make it challenging for him to get a foothold in relationships, a community, a job, or develop recreational interests. 'I have this sense of a terrible lack of traction across the board.' I didn't know him well, but I had always found him angular, aloof-seeming and jumpy, his eyes always set to large like a nocturnal animal. Ruben had been working with his therapist with slow, incremental progress for several months – which was remarkable, as he had never been able to attach to a therapist before – when she suggested they try some MDMA-assisted sessions. For these they were joined by the psychiatrist who Ruben's therapist had been under for ten years as a patient; they had since become business partners. 'I loved the idea of male and female energies,' Ruben told me.

He described the sessions, which cost $3,000 each, in detail. He and his wife would rent an Airbnb and drive to the psychiatrist's early the next day. He would take the MDMA in a capsule lying on a large double bed in the middle of comfortable, living-room-like space (the psychiatrist's treatment room), with the therapist and psychiatrist seated either side. As Ruben 'came up' on MDMA in his first session the psychiatrist began to cry. Ruben's childhood had been populated with aggressive, bullying male figures, so in his mind the psychiatrist's tears were part of the transference, signifying a depth of feeling for his patient. 'I thought, he is one of the most empathic man I've ever met.'

The psychiatrist turned to the therapist and addressed her. 'He's been alone for so long.'

'I'm looking at his eyes. They are unreaching and unreachable,' the therapist replied to the psychiatrist. The two professionals were having a conversation in the third person about their patient.

'He doesn't care if you hear him. He's not interested in your response,' said the psychiatrist.

The psychiatrist was right, according to Ruben. 'I wasn't there . . . Like I had retracted deep into myself. I had no nerve endings. Then I felt like I was getting sucked into a black hole. It started squeezing my insides – it was really horrible, making me contort.'

'If the black hole could speak, what would it say?' asked the psychiatrist.

'Fuck you.'

'Tell it to her, then.'

Obediently, Ruben started yelling at his therapist. 'It didn't feel at all right when I started, but then it was amazing to have her receive it. She didn't recoil, she didn't wince, in fact she giggled, feeling the release of my tension.'

For the second session Ruben arrived full of apprehension. 'I took the MDMA like the first time and lay in silence on the bed, waiting. The psychiatrist was nervous, agitated even, as he addressed the therapist: "We are nowhere, *he* is nowhere, we are just floating, totally untethered, I don't know what's going on here." '

'I'm here too,' the therapist told Ruben. Ruben knew what she meant: that she was right where he was psychologically, that she knew the same emotional place that he inhabited. (Her own complex history of abandonment had been the focus of her decade-long work as the psychiatrist's patient.)

But the psychiatrist didn't appear to like the identification between the other two people in the room. 'Have you been hiding this from me all these years?' said the psychiatrist to the therapist.

'It's like he's jealous,' Ruben explained to me; 'that she's having this moment of profound self-discovery with me rather than him.' Moments later the psychiatrist got up and walked out.

His absence allowed Ruben to experience the drug's effects without the more complex dynamics. 'I could feel the effect of the MDMA, waves of pleasure. I let my body do whatever it wanted to, spontaneously, lolling my head, kicking my legs. I was looking for my therapist's eyes, I felt so hungry for her eyes . . . We looked at each together for a while . . . I suddenly felt this deep hunger to be seen.'

I prompted Ruben about what he thought about the psychiatrist's behaviour: his tone, his eccentric reactions, his 'charisma', which seemed to me to have been fuelled by the psychiatrist's own neuroses rather than the patient's well-being. But at that point, in a manner that was typical of therapeutic power asymmetries and therefore the potential for abuse, Ruben explained that he still trusted the psychiatrist: his intuition, his creativity, his empathy.

That had changed by the time I received a letter from Ruben a few weeks later, shortly after arriving in Oregon.

I have some big news. My therapist has cut all personal and professional ties with the psychiatrist who facilitated my MDMA sessions with her . . . Evidently, he said something to her in conversation about an incident years ago where he clearly violated a client. She challenged him on it and, instead of expressing remorse or admitting wrongdoing, he justified his actions as being for the client's good. She then went to someone who knows this psychiatrist well in his work, who filled her in on other such incidents with other clients, up to the present day. I don't have many details and I want to be discreet, but it's serious enough that she is taking legal action. Needless to say, I did not have my scheduled session this past weekend.

This is a very complicated situation because my therapist had seen this psychiatrist as a (non-psychedelic) client for many years, and says that she never had any indication of what she now knows about him. Two years after she stopped being his client, he introduced her to psychedelics, guiding her with MDMA and mushrooms before inviting her to partner with him as a guide herself. To complicate matters for people like me, up until this current crisis, she held this man in incredibly high esteem, trusting him thoroughly. And because I trusted her and her judgment, I inherited her trust of him.

Curiously, I didn't start to seriously doubt his role in my sessions until I spoke with you, questioning my previous assumptions as the words were coming out of my mouth. Where his harmful behaviour intersected with me is that he evidently has a pattern of cultivating various forms of dependency in some of his clients, and he seemed to be grooming me and my therapist for a similar kind of relationship – I as a kind of dependent child to her as a mother-figure – all in the name of healing from attachment trauma. And even though my therapist and I were self-correcting away from that paradigm, he generated a lot of confusion and certainly touched on some of my deepest vulnerabilities. Still, I don't regret the sessions at all . . . Next time I will definitely have my bullshit detector off mute.

As you can imagine, this has been shattering for my therapist, but she seems to be handling it better than I would have anticipated . . . She is getting help from, among other sources, a psychologist who specialises in psychedelic abuse. She's learning a lot and passing on some of what she's learning to me, about the dynamics of abuse in a therapeutic context, and deconstructing some of the psychiatrist's influence on our work, such as putting so much emphasis on the client's attachment process with the therapist and where that becomes disempowering.

Trauma depends on the catastrophic violation of trust, the fracturing of one's faith in the safety of the world. We know that abuse happens reliably, predictably, even in the 'securest' settings – families, hospitals, therapy sessions – and the media give disproportionate emphasis to the reporting of such cases. Our 'surprise' at them is therefore a facet of our collective capacity for amnesia or denial, or as Freud would say, our belief in 'civilisation', despite the rapacious impulses and biases which have always besieged us.

The mechanics of trauma, abuse and different kinds of defence mechanisms are at the heart of the psychoanalytic tradition. The dynamic between therapist and patient is both the subject of therapy and the mechanism by which it seeks to bring about change. Modern science regards psychoanalysis as gratuitously speculative and ideological, which means that the some of its most profound insights on the dynamics between clinician and patient, the creation of 'observer effects', the importance of reflexivity – blind spots in much of scientific methodology – are also neglected.

Being more of the world of science than of psychoanalysis, psychedelic-assisted therapy as it exists today largely ignores these insights as well, and depends instead on a rudimentary model amounting to not much more than basic good-sense counselling skills. And maybe that is all that is required: the 'medicines' themselves are often framed as 'the therapist', while the therapists become 'facilitators'. But this was not always the case. In so-called 'psycholitic' therapy of the 1950s, patients would take small to moderate doses, typically of LSD, over the course of multiple sessions led by analytically minded guides, to find the heart of traumatic experience. In the much-celebrated work of Stanislav Grof this meant going past the moment of birth (as with the hour of one's death, this 'watershed' was commonly encountered on psychedelics) to perinatal challenges and problematic past lives. By contrast, the current form of psychedelic-assisted therapy, at least as presented in the clinical literature, takes little interest in parsing the relationship of the trip to that of the patient's life more generally. At present the sensibility of psychedelic-assisted therapy remains wilfully small.

Kate, the MDMA therapist, was a bookish-looking former architect from the East Coast. At least to the waist she was. From there on down she became, like a centaur, something different; her striped Indian pants and barefoot shoes those of a contact impro yogi. While interviewing me the architect was at the fore; setting rules, keeping records, making sure everything was in the right place; propriety, precision, deliberateness. I felt myself relaxing, attaching to her authority and orderliness. She spoke quickly, with the haughty beauty of her namesake Katharine Hepburn.

We were talking in the well-appointed Japanese-style gazebo at the bottom of her garden in north-east Portland, under the occasional eye of a former husband who lived in the house next door. A precaution, she explained, as she was less than a hundred pounds, and some of the people she saw were veterans three times her size, with arms thicker than her neck. (She would still have inadequate security, I thought, if they tried to take back an imagined Taliban stronghold during the session.) Business was booming: she was seeing two patients a week at present, switching out her normal cranio-osteopathy appointments. Each patient required a nine-hour treatment session, and additional sessions for screening, intention-formation and integration. She had a waiting list that was several months long and included 'illegal' referrals from private psychiatrists. What kind of patients were the psychiatrists referring I wondered? Were psychedelics destined to become the latest way the profession passes on the untreatable, a Hail Mary, the treatment that comes after the official treatment of the last resort?

Kate spoke about her 'lineage', meaning her informal underground training, which was not with MAPS. MAPS had an increasingly bad reputation as far as Kate was concerned: there were the ongoing legal cases following allegations of abuse, and the training was 'overly formulaic', 'thin' in terms of its understanding, and expensive. Kate explained that MAPS had struggled to raise the finance for the final stretch of its Phase 3 trial and, in keeping with general market tendencies, was having to look beyond philanthropists to private investment. In MAPS' case this meant 'crypto bros'. A quick Google search showed that in June 2022, a couple of months after my MDMA session, Christie's, the world-renowned auctioneers, were proud to present 'Cartography of the Mind: a Curated NFT Sale to Benefit the Multidisciplinary Association for Psychedelic Studies (MAPS) in collaboration with Ryan Zurrer, founder of Dialectic and Vine Ventures'. Though the Psychedelic Renaissance has barely recognised the arts and disciplines beyond clinical science, it's not coy about allowing psychedelic art to fund its needs.

Kate, meanwhile, was being cryptic about who exactly had mentored her, but I inferred – or perhaps, to use a technical term, I fantasised – that she had learned at the feet of Anna Shulgin, the recently deceased wife of Sasha, who had pioneered MDMA-assisted therapy for trauma and attachment. The fantasy made me attach to Kate all the more strongly.

Of all the expertise, all the millions of words I had slogged through in the psychedelic field, Sasha Shulgin's double magnum opus *Pihkal* and *Tihkal*[12] (names evoking images of Amerindian deities from ancient rites, but really acronyms for 'Phenethylamines . . .' and

'Tryptamines I Have Known And Loved') stands out like birdsong at the end of war: the astonishing distillation of chemical expertise, decanted in pellucid prose with sublimating humour, the portrait of a life compounding science and bohemianism and, more intimately, the purity of the bond and enduring attraction between husband and wife (the book's subtitle is 'a chemical love'). This is what real expertise looks like, I couldn't help but feel as I read it: the same first-person expertise of the nineteenth century, poetically inclined toxicologists who applied venomous spiders to their arms and then wrote the results down to the point of death, their handwriting increasingly illegible or *spidery* as the toxins approached lethality.

The book's format reminded me of chemistry homework at high school – the essays sectioned according to Synthesis, Duration, Dosage, Qualitative Comments and Extensions and Commentary – only this was the nonpareil of chemistry homeworks: a thirty-year opus in a shed at the bottom of his garden, with FDA approval because the compounds were so novel, literally fresh as a new-born test-tube baby, they couldn't legally be banned (until they were on a whim), and the only type of homework you can legitimately excuse yourself by eating.

Shulgin is generous enough to leaven what is by turns meticulously procedural and esoteric with wry, humane treats. These take a range of forms, including philosophical speculation:

> With an 80-carbon side-chain, would one-thousandth of a single molecule be enough for a person? Or might a single molecule intoxicate a thousand people? And how long a chain on the alpha-position might be sufficient that, by merely writing down the structure on a piece of paper, you would get high? Maybe just conceiving the structure in your mind would do it. That is, after all, the way of homeopathy.[13]

. . . Nerdish humour: 'There is a jingle heard occasionally in chemical circles, concerning the homologues of methyl. It goes, "There's ethyl and propyl, but butyl is futile." '

. . . A serial killer's dispassion: 'A fourth mouse, at 150 mg/Kg (ip), entered into spontaneous convulsions within 10 minutes, and expired in what looked like an uncomfortable death at 22 minutes following injection. What was learned?'

. . . And all-round professional delinquency. Like Frank, Shulgin lived and died by his 'drive test' for different concoctions: 'Driving was easy.' 'Driving was delightful.' 'Could I drive, I suspect so?' 'Driving would be impossible.' 'The white lane-marker stripes were zipping up . . . like disturbed fruit flies leaving an over-ripe peach.' The drive test was offset by the safer, more stolid, bourgeois 'museum test': 'The Rodin sculptures were very personal and not terribly subtle.'[14]

Unlike the norms of much of contemporary psychedelic science, Shulgin is honest about the limits of expertise, and of accidents, whether chemical or existential. 'I had simply picked up the wrong vial. And my death was to be a consequence of a totally stupid mistake . . . A person may believe that he has prepared himself for his own death, but when the moment comes, he is completely alone, and totally unprepared. Why now? Why me?'[15]

Shulgin's work is that of an authentic Renaissance man: someone steeped in chemistry, but whose intellectual depth and breadth is saturated with the broader culture. The science, pioneering in itself, is combined with the psychonaut's first-person experience, knowing the new drugs molecularly, but also as experiences; intimately, that is, as wonders, a gourmand's delicacies – demons, delinquents, tricksters, teachers – bringing together the bohemian the scientific and the surreal. The good life, in other words. This is what engaging with psychedelics should be, I felt as I read him.

The architect had gone, the centaur's New Age bottom half was in control now. Kate stood over me holding an eagle feather in one hand (there must be a lot of semi-denuded birds around psychedelic retreat centres), burning sage in the other. She made what was to become, over the following weeks, familiar invocations to the Spirits of the Four Corners: heaven, earth 'Pachamama', the Underworld and, the one I always resonated with, mid-air, no man's land, the in-between.

When I returned, as requested, her 'sage-ing' of me, smoking her from head to foot, I was moved by the sight of a large hole on the underside of her 'fun' striped socks. I wanted to repair them. I wanted to look after her. As though sensing an attachment fantasy she handed me an eye patch. 'I don't want you distracting yourself from the depths.' I thought of Ruben, hungry for his therapist's eyes.

The day before, when we had spoken generally about her therapy business, Kate told me that in the last few weeks four of her patients had *discovered* traumatic abuse as children during their journeys. From clinical experience I knew something of the complexity of a patient realising a past trauma in a therapy session. There were certain signs and 'intuitions' that one could carefully unpack further either to help establish the authenticity of the information or to formulate it as a symptom of something else. This had happened to me twice in fifteen years, and on both occasions I had floundered in my lack of expertise and had to seek specialist supervision.

Kate was encountering them at astonishing frequency, and outside of a supervised clinical framework. The revelations came not within an enduring relationship between therapist and patient, but as one-offs, unsolicited, under the influence of MDMA, a drug known to enhance suggestibility. From later conversations with other psychedelic facilitators I understood the 'recovering' of 'traumatic' memories was commonplace across different drugs and settings.

Testing the reality of such experiences is complicated. The patients Kate saw were not a random sample of the population, but largely self-selecting. They were likely to know something about the association of MDMA with trauma before their experience – it was hard to avoid in the public discourse – so there were already in place so-called placebo enhancements. It's possible and statistically likely that the 'discovery' of specific traumatic origins includes false positives: discoveries that work symbolically as a neat, bow-wrapped origin story for all the patient's inchoate symptoms, to everything that's otherwise wrong in their life, but which do not correspond to any actual event. There is also the more egregious danger of false negatives: that the patient's real discovery is taken metaphorically. But the frequency of these revelations in Kate's practice suggested a tendency to take whatever the trip 'revealed' literally.

Kate was evidently a skilful and conscientious practitioner. She insisted her patients were in an ongoing relationship with their own therapist, and that confirming or disconfirming traumatic discoveries was beyond the scope of a one-hour 'integration' session. The problem of taking the trip literally was not hers; rather it was endemic to the broader culture. If psychedelics are 'therapeutic', and therapy is viewed through the lens of trauma, then trauma is more likely to be seen. While psychoanalytic theory might offer useful ways to expose or highlight the symbolic nature of such discoveries – as primitive defence mechanisms such as 'projection' or 'splitting' – or to point out the ways in which the therapist can unwittingly help to shape the narrative of a 'revelation', even with micro-nods or whispered 'uh-huh's, intended as tokens of empathy or containment, it remains out of favour with most practitioners.

Twenty minutes in I'm in familiar psychedelic territory: the same automatic stress response – pins and needles in my face, the urgent wish to pee, blood turned four parts cortisol, battle-ready, waiting for a trauma whose origins were in a remote past. My hippocampus decides to serve up the memory of the night with Aurora where I was finally given the *gift* of understanding madness, not from clinical textbooks, but from a first-person perspective. I tell my amygdala to tell my racing heart to be quiet. *It's nothing*, as one might say to a child before an impending catastrophe.

'Aunty Em doesn't want to overwhelm you,' says Kate.

'Aunty Em? From *The Wizard of Oz*?'

'That's right.'

The film always terrified me – the hideous green witch, the talking trees, the flying monkeys . . . the archetypal bad trip. I need to make a run for the gazebo door, except: this beating heart, this tingling skin, the sound of high Munchkin-like voices ringing in my ears – it's not fear, it's bliss. Unlike most psychedelics MDMA excites the dopamine as well as seritonergic systems. I felt liquid euphoria, each drop landing on my raw, catheterised heart. The need to run away melted. *Medicine*. There was no other word for it.

'There's a deep cave of the heart,' says Kate.

From under the eye patch Kate has *become* Katharine Hepburn at her gentlest, in a nearby rocking chair knitting a shawl, which makes me Spencer Tracy at his goofiest. It's official: I am attaching.

> I have never felt so great, or believed this to be possible. The cleanliness, clarity, and marvellous feeling of solid inner strength . . . I felt like 'a citizen of the universe'. Everyone must get to experience a profound state like this. I feel totally peaceful. I have lived all my life to get here, and I feel I have come home. I am complete.[16]

Shulgin on 120mg. After my third bump, taking me to 160mg, I would think, 'Right, Sasha, but not right enough. I would go further.' I was bodiless, without edge. For a moment I mistook the place where I found myself for an abyss, the locus classicus of every possible trauma, but as each drop of heavenly nectar landed I let in another possibility: this was perfect, as Shulgin had it, but less abstract. I had arrived at a very specific place – the place where every need is met, immediately, without being asked for, where everything is still possible because nothing has happened yet. Total provision. Perfect attunement. I was inside my mother, waiting to be born. Intra-uterine bliss, just as Grof's regressive psycholitic therapy predicted. The healing was in the realisation that everything was already whole.

The researcher Matthew Baggott advocates for a more clinically nuanced understanding of psychedelic experiences. As well as being an 'empathogen', MDMA can be anxiogenic, producing anxiety at certain doses. But, Baggott suggests, it allows the person to be in a relationship to a negative emotion while also being relaxed, and therefore less responsive to fear's entailments.[17] My need to run was the classic 'masculine' response to anxious physiology – fight or flight. But the oxytocin released by MDMA allows for a more

'feminised' response to fear: a response that is deeply receptive, permitting intimacy, craving relatedness.

There's a scar on my leg which is forty years old, from an operation to remove a cancerous bone growth from my tibia. For two
weeks I was on a children's ward with half a dozen others from
unknown, exotic parts of the city: a boy in traction for hips shattered when he'd written off his dirt bike after sniffing glue; a
twelve-year-old who'd had his stomach pumped after drinking
his mother's vodka; a girl with a double-barrelled name who was
undergoing chemotherapy; a tall Jamaican boy who had bandages
round both wrists and only made eye contact with a Rubik's cube.
Strangers in a strange situation. Over that fortnight an unlikely
gang had formed. I loved everything about being in hospital, even
the smell; perhaps it's why I ended up working in one. I didn't want
to go home. On the night before discharge, my bags packed, I threw
myself out of bed and re-opened the wound on my leg. The scar I
was left with was, following nerve damage, a mix of dead and over-
sensitive. I never let anyone near it, wouldn't touch it myself.

Now Katharine Hepburn was massaging coconut butter into that
same scar while I, my heart in my mouth, tolerated it, astounded
that anything could break down defences erected all those years
ago, show them to be long outgrown. For those few hours I felt
with astonishing clarity that the truth of our childhood is stored in
our bodies, how this miraculous medicine seems to crack open
trauma's vault. I never wanted to leave Katharine's side.

People told me it would be like five years of therapy in one day.
That's not been my experience . . . Overall it's been so hard. My
adaptive strategies are breaking down, the more that happens the
more inept I feel. There's been sessions where I can't even talk,
and afterwards I don't know how to show up at all. It's the opposite of how I would hope things would go. But counter-intuitively
it might be the way forward, a grinding down of self rather than
breakthrough . . . It's fucking hard . . . definitely not the rocket
ship to healing.

Ruben's words, after five further guided MDMA sessions without the psychiatrist. The trauma patient arrives at the dead, insensate place, the bedrock of his emotional geology. *Can you love that? Can you inject it with feeling and care for it?* The research suggests that, like many of the psychedelics, MDMA is a catalyst for change, but it also suggests that such change is often not the explosion of sudden revelation but long, incremental and hard-fought. Some of the respondents on MAPS' MDMA Phase 3 trials had fifty hours of therapy alongside their journeys, suggesting that talking is as at least as critical as chemistry.

We eat clear chicken soup together across a small table, Kate and I. It's a little uncomfortable with the eye patch off. In the days to come, Nigel, my therapist, will tell me that something about me seems different. 'Like something has been undone, and something else has taken its place.' I will tell him about Katharine Hepburn, who gave her touch with its special voltage, delicate one moment, forceful the next; this woman who— And I interrupt myself to tearfully say, 'I need to buy her new socks, Nigel!' But on that evening, it is just awkward, as though too much has been given away that was not earned. Spencer and Katharine have gone; just the sound of the spoons chinking against the bowls and hitting our teeth.

4

The Substitute Trip

The Pacific paused a moment before tearing into Highway 1, shifting this bit of the continent east, a bite at a time. The Canadian border was only a few hundred miles north, and I was on my way to experience the psychedelic hype-machine's motherlode, its pay-per-view main event: *The most powerful entheogen known to man. The strongest psychoactive molecule on the face of the planet . . .* Ibogaine could be had along the same highway, more or less legally either side of the US, in British Columbia to the north or across the southern border in Mexico, where various pit-stop shops had sprung up in the last years to cater for the treatment of the American opiate epidemic.

It was a road trip; things fell into place *because* there was no plan. Some more astrologically minded navigators might claim it was not mere coincidence that all the major psychedelic compounds were arranging themselves up and down the western coast of the American continent. But that kind of thinking – of psychedelic ley lines, of biorhythmic planes, of Gaia's longitudinal truth, as though a secret ventricle in the world's brain had opened – was, like me, a long way from home, and would require a tectonic shift in my world view, when up to this point in my journeys I'd experienced only the mildest of continental drifts. Like most theories of this nature, it also required ignoring certain material facts: that Oaxaca State, where magic mushrooms are indigenously used, wasn't so much below British Columbia as east of Chicago; that the Amazon, where ayahuasca originates, is nowhere near the Pacific coast; and, most killingly, that *iboga*, from which ibogaine is naturally synthesised, is found indigenously 8,000 miles to the east in West Africa (which was why I had elected to take the easy-to-access synthetic analogue rather than the real thing). But what this New Age cartography lost in geographical accuracy, it gained in paranoid best-fit: Silicon Valley – the dark, algorithmic heart of mind control, of 'Cyberdelia' – fell on the exact same plane. The West Coast setting was also really a *set*, an ideology, a unique mix of experiential libertarianism, entrepreneurialism and techno-utopianism, as one commentator, I. Hartogsohn, put it.[1] And there was a prophetic voice in the Valley I had been keen to talk to.

So that when that voice invited me for tea over the phone on the same morning I was driving up I-5 bound for British Columbia, it was enough for me to turn the car around and make a 700-mile detour south.

'Timothy Leary was a friend; someone I knew a long time – we got up to hare-brained adventures back in the Nineties. Like the time I had to get him out of a talk he didn't want to do at Esalen, so we found this impersonator, a look-alike, and snuck the real Tim out in the trunk of a car.'

Jaron Lanier, father of virtual reality slash computer scientist slash philosopher slash technology ethicist slash Octopus (Office of the Chief Technology Officer Prime Unifying Scientist) at Microsoft slash composer slash musician slash musicologist slash Islam economist (the list goes on), all-round Renaissance Man, and not particularly partial to the product. 'I was never really into psychedelics – they just didn't do it for me.'

I'm meeting him at his home in Santa Cruz while he takes a break from one of three books he's writing simultaneously (one on consciousness and two on music), the second of which is written in a prose analogue of a musical fugue – 'I like to set myself *little* challenges, keeps me on track. You wanna know the secret of productivity? Cross-procrastination.' We're sitting on the bottom section of a high-ceilinged, split-level atrium, the upper part of which is a pantechnicon of musical instruments, seventy at least, collected from round the world, as though waiting for a cross-cultural orchestra to arrive. (*Most* of his instruments, Jaron tells me, are at another home in Berkeley.) Looking at them evokes a kind of uncanny valley – the eerie response to the nearly real – or a neuro-psychological test of object recognition: they are formally unmistakable as musical instruments, and yet I've never seen most of them before. By this point in my research I'm framing most things in psychedelic terms, and the unsettling amalgam of the familiar and the never-seen-before is beginning to feel like a signature feature of the trips themselves.

For the most part Lanier speaks fondly of Timothy Leary, the Sixties voice of psychedelics who's either 'a hero of American consciousness' or 'the most dangerous man in America', depending on who you're asking (Allen Ginsberg or Richard Nixon). Lanier's word for him is 'meta-syncretic': a man who took his own notions of set and setting and collapsed them into himself, made them an art form, so that, for example, a good musical experience had to do with a good political experience had to do with a good interpersonal experience. If I was beginning to see things in psychedelic terms, Leary could *only* see them in that way. 'Everything for Tim was not only merely interconnected but different facets of the same thing.'

This was part of Leary's charisma and charm, Lanier goes on, but it could become annoying when it reduced the Harvard psychologist's capacity for critical thinking. 'When you become too loose with your associations, then low-grade aspects of human nature can come in and get a foothold when they shouldn't.' In Leary's case, he tended to succumb to self-promotion, to glamour, to hype. Lanier recalls waking up one morning, shortly after meeting Leary for the first time, to discover his own face on the front of the *Wall Street Journal* and *New York Times*. 'Is this crazy freak giving us electronic LSD?' ran the headline. That was Leary's coinage for virtual reality. He had a knack for arresting coinages. 'While I might have looked the part' – thirty years on, Lanier retains copper-coloured dreadlocks that reach the floor – 'the reality was much more square: I was spending all day every day making surgical simulators to assist the training of physicians. I don't even like acid.'

Whereas today's boosterish scientists keep themselves fenced off from the counterculture, Leary, for good and for ill, was a swinger. 'He had all these people hanging around him with crazy schemes to build a VR system for dolphins to wear, or devices that could put your brain in orbit . . . It seemed like everyone in the entourage had already tried all the psychedelics out there – scrambled to the top of undocumented mountains, or the farthest reaches of the Amazon to find these obscure plants, or hung out in Sasha Shulgin's basement waiting for the latest compound to cook – and now they wanted VR. So I'd hand them a headset and say, "Well, here it is. Have at it . . ." They'd always be disappointed.'

Besides listening to stories of the old days, I wanted to speak with Lanier because researchers were beginning to test high-resolution VR (virtual reality) as a possible moderator or even substitute for the drugs themselves in psychedelic-assisted therapy. Clinical trials were being run on depressed patients to compare its efficacy with that of psilocybin and ketamine. In a trial of the capacity of a new VR ('Isness-Distributed') to generate 'intersubjective self-transcendent experiences (STEs)' (as measured by the same instrument that Johns Hopkins used to measure mystical experiences), the results were 'statistically indistinguishable' from psychedelically induced STEs. For those who might be fearful or disinclined to take a substance, this technology could make therapeutic 'trips' drug-free, just as the pseudodelic (psychoplastogen) researchers were trying to make therapeutic drugs trip-free.

From Lanier's perspective, VR is a different order of accomplishment to most digital technology. He points out that by creating successful illusions – generating digital images, worlds, even people that 'seem more or less like reality' – 'most technology gravitates towards treating all experience as some sort of illusion'. Its implication is that everything, ultimately, can be digitally rendered, faked. And because we absorb that narrative, we 'end up thinking that reality only *seems* like reality, and people only seem like people, and the whole thing's an illusion and so on . . .' Whereas VR rotates this in the other direction, flipping our intuitions, as he explains. 'When we're in the midst of a VR experience we can't help noticing there's something real and extraordinary going on.'

It made me think of the nuanced relationship psychedelics can have with actual reality: at one moment the drugs appear to 'create' certain experiences, the next they suggest that the things being experienced are already 'out there' in the world, albeit in less manifest form. They pose the question: are we 'creating' these experiences, or are they visited on us by material changes in brain chemistry that allow us to perceive reality more fully, or by the activation of something like the Freudian unconscious, or even, as some believe, by the promptings of external supernatural forces?

Meanwhile Lanier was speaking of the parallels between VR and psychedelics. He believes that the experience one has in VR depends, like psychedelics, on priors which inform notions of set and setting. 'In both one has to take in the surrounding aesthetics and politics, culture and philosophy, which govern how meaning is made.' This level of sensitivity to their 'embeddedness', Lanier agrees, is usually missing from thinking about psychedelics. It makes him circumspect about naïve talk of the 'therapeutic' potential of psychedelics or any kind of technology. 'To the degree to which you can treat people using VR or psychedelics, you can also potentially manipulate people – in the way they are already being manipulated using TikTok and Metaverse, realities that are tremendously dark and frightening. It means I can't enjoy the technological success of VR innovation or streamlined psychedelics in a straightforward way.'

One notable feature of our conversation: each time Lanier is drawn into comparing VR with psychedelics, or anything with anything else for that matter, he's also inclined to punctuate their distinctiveness, which gets folded back into a different level of comparison. 'VR is a form of expression, like writing or music, that becomes more intense the more expressive somebody is. Psychedelics are intrinsically more intense than VR, historically at least, possessing infinite resolution, but they're also more subjective. VR is fundamentally acting on stimulus' – by which I assume he means triggering the visual system in the first place – 'whereas the drugs are acting chemically on different layers of perception simultaneously (although there's variation there too). That aside, [with psychedelics] we are dealing with something that's more an internal experience than an external one, whereas with VR – even with the most advanced systems we might foresee – there's something *shoddy* to the experience, an intrinsically low-res view of the world.'

This leads Lanier into a favourite topic of his: whether VR could ever get so good that you can't tell the difference between it and reality. In his view, it's unlikely, not because the VR won't get better, but because people's ability to distinguish the fake from the real will evolve with it. Think of the geometric progression in special effects from Hitchcock's *The Birds* to Cameron's *Terminator* and then *Avatar*, with each generation seeing the realism of the previous as 'clunky' or 'unrealistic'. As Lanier says, 'The idea of *realness* is itself a moving target. People never stay fooled; they get more sophisticated.'

I would also add that being fooled is determined not just by something's realness or otherwise. It also depends on our motivation. Sometimes we may want to believe things that are less than real. In the same way one wonders if 'psychedelic' is also a moving target; if in drug development's near future people will become so habituated to current iterations – de-sensitised by over-use, or by their intensive engagement with all manner of high-res technological illusions – that they might need more powerful, more original, stronger medicine, third, fourth and fifth generations to receive any therapeutic benefit. (Or, even if the habituation itself isn't real, whether the need to update will be driven by market forces). And, of course, this will be co-terminous with future movements in psychedelic nostalgia for the good old days of rough and ready second-generation compounds. Equally, if the technology for direct brain stimulation progresses, it's conceivable that it, a form of VR, might accomplish identical resolution to psychedelics. Elon Musk and others are working on a 'layer' of 'digital quantum processing' that 'interfaces' with the brain's cortex in a far more complex form than anything seen to date. Lanier agrees that this will have both attractive and unattractive ramifications. 'There's likely to be commercial and governmental involvement that can create conditional access that might have some kind of expression of power built into it.'

I find myself admiring how the ethicist in Lanier is always lightly holding in check his more expansive, even rhapsodic inclinations toward digital technology. These moral and political dimensions of his thinking feed back into a further comparison of VR with psychedelics. 'Both can have a kind of quality that allows those in control of the distribution system to smuggle in an unacknowledged hack. In the case of VR it might be Meta insisting you buy into their commercial model' – which Lanier was one of the first to diagnose as a form of behavioural manipulation – 'and that's pretty awful. On the other hand, with the distribution of psychedelics there have been all manner of unsavoury characters. The way the thing is being commercialised now makes it very hard to organise in such a way that doesn't invite some kind of corruption eventually. I'm not saying it's impossible, but even when you start with the best of intentions you might fall short.'

According to the philosopher William McCaskill, we are at a unique moment in the history of technological development; its acceleration is greater both than anything that has preceded it and, most probably, anything to come.[2] It puts us on the brink of various 'singularities' in AI (which is mathematics for radical and irreversible misbehaviour): in the creation of machines with general intelligence, most obviously, but also within medical science, and the possibility of high-spec synthetic psychedelics that may outstrip our capacity to catch up with the significance of the changes. For McCaskill the technological advancements have come before we have developed sufficient moral capacities to contain or safeguard them.

Lanier shares such ethical concerns. 'The same technological philosophy, materialist or reductionist, that's being used to construct the experience of the metaverse, is used to understand all of our experiences, including the science of psychedelic experience. Digital technology has a way of reducing the [variety] of experience.' He describes this as a collapsing of all kinds of experience into one kind of experience: the materialist one. 'For a real nerd, there's a collapse of experience itself. Everything turns to nothing. That's what I'm trying to gravitate against.'

And just as I think he's finished, at a point that suits my argument – meaning the precious quality of experience itself, which is under threat in psychedelic science – Lanier jags the other way. 'The idea of reductiveness is not necessarily problematic: we have limited notes in music and words in language – the piano is just buttons, and we press them in order to make a Bach fugue, and somehow we escape reductivism . . .'

Lanier, like the logical implications of anything in a real-world setting, is never quite finished. By the end of the conversation, I felt unconstrained; talking with him made everything seemed connected in ways that were existentially exciting and troubling, morally ambitious and deeply corruptible. But then, by way of simplicity, Lanier told me that for him the most important experience of all is the moment when you take the headset off, or the trip is done. 'And there is reality all over again, only now it's so fragile, liquid, precious.' He lets the suggestiveness of that settle for just a moment. 'Of course, the market corrupts that too, and now they're packaged as "palate fresheners" or "re-setters".' As if reality is a tonic that can be sold back to us with psychedelics.

Curating an ibogaine experience in November 2022 remained at the low-tech end of the psychedelic business, but that didn't make it any less corrupt. The opposite, in fact. It was rather like being on the trail of the Maltese Falcon: underground hearsay, international extortion, undocumented deaths, peopled with dodgy characters with big, wide Peter Lorre eyes saying things like, 'Can I level with you?' It started with me watching a feature-length documentary about a young, female heroin user's efforts to clean up using informal psychedelic therapy in British Columbia.

In fact, that's a lie, and I haven't got to the first sentence of my confession. Lying and addiction are siblings, and I want to be straight with you.

I was an addict. For a little over a decade I couldn't stop drinking and taking hard drugs. Alongside psychotherapy and Twelve Step groups, part of my rehabilitation was vocational: I helped establish an NGO to serve the practical and emotional needs of the street homeless in the north of England: younger heroin/crack cocaine addicts and older street drinkers. I knew first-hand something of the 'bio-psycho-social' complexity that causes and maintains addiction, and the poor efficacy of any pharmacological treatment of addiction.

It meant the world of the documentary was familiar to me. I also found the film provoking in ways that were not intended. At times I warmed to the girl: her pluck, her tricksiness, her frankness. How, when asked how she was feeling by a supporter after a first treatment with mushrooms, she replied, 'Honestly? I just love being off my face.' A refreshing take, not commonly admitted in the psychedelic literature. In contrast to the film-maker, a desperately concerned friend, and his constructed naïvety about his protagonist, I spotted her serial mendacity and appreciated its authenticity. 'Can I be honest with you?' she kept saying, after another relapse. Then again, twenty minutes later: 'Can I be *really* honest with you?' Then, following the film-maker's tetchy, 'personalising' disappointment, she pulls the addict's Houdini routine and reproaches him for not seeing this continuous updating of 'honest' as symptomatic of her *disease*.

Other things, however, got in the way of my sympathy: the lack of 'art' to the storytelling, the earnestness, the soft piano music and, most of all, the fatuous equation of addiction with trauma. At one point the girls says she *feels* as though she has PTSD, but can't remember what caused it. No matter, the healers tell her, it's there, the memories are there, you just haven't recalled them yet. But her childhood was 'sweet', she tells them. Her mum and dad were kind. Never mind, she's told, the addiction carries a deeper intuition, a canary singing in the mine of her early development. That's why she needs the medicine of ibogaine: to 'regress' her, connect her with the truth from which she has been dislocated.

It seemed not just fatuous but dangerous; as though the saddling together of two poorly understood concepts, trauma and addiction, in a causal sequence, might somehow make both better understood. Then an exotic psychedelic is added to the triangle: it may cure addiction (as well as other diagnoses) by revealing trauma, it may cure trauma (as well as other causes of mental illness) without itself being addictive; occasionally it might *cause* trauma or lead to a kind of psychological dependence (which may be classified as addiction), but it's hard to market or brand that degree of conceptual overlap, so things are kept artificially simple: *ibogaine treats addiction.*

After the Silicon Valley detour I was heading north again, up I-5 to Clean, a company specialising in ibogaine treatment that operated out of Salt Spring Island, British Columbia. John, 'Mr Clean', had stolen the show in the documentary as a sharp-dressed, sharp-mouthed, no-nonsense *geezer*, an Englishman in British Columbia: herringbone jacket, boxer on a lead, like he'd taken a wrong turn in a Guy Ritchie movie. He was just the same in real life. I stopped off again in Portland to speak to him on Zoom, a conversation which began with me asking how he ended up a film star.

'What's an ex-junkie doing anywhere?' he replies. 'One minute I'm off my box at Stonehenge in 1987, the next it's 2018 and I'm getting smackheads well in the Pacific North-west.'

Stonehenge because he was a Traveller, a Hawkwind fan who liked acid before he loved heroin, and because he enjoyed scrapping with the Old Bill, who saw 'nothing as more important than preserving law and order around pagan monoliths'. Behind him an A-frame window looks onto rolling hills covered with tall cedars and conifers, and between them glimpses of the ocean. Clean move from rustic Airbnb to Airbnb, according to their bookings. From time to time a middle-aged woman, Bella, a registered nurse and administrator, hands him something. Basic medical care is required in ibogaine treatments, because of health risks to certain groups of people that make it more dangerous than other psychedelics. While there have been few controlled trials of ibogaine, case reports have described sudden death, prolonged cardiac arrest, arrhythmias, liver problems and seizures after use, especially in the medically vulnerable. As opioid users are more susceptible to cardiac problems, pre-treatment electrocardiograms (ECGs) and blood tests are standard.

John declared himself well disposed to the medical model, offering, I imagined, a measured counterbalance to what sounded like the chaos of his former life. 'I'm a clinician, not a shaman. I supervise pouring this medicine into the arms of fucked-up junkies to get them well. Smudging patients' foreheads with ash is a about as wu-wu as I get . . . I mean, sometimes we stick on some African tribal music from Spotify if they really want it, but personally speaking I can't stand the racket. Brings out my inner Victor Meldrew.'

For all his patter he was fluent in the language of the clinical technician, explaining that once the bloods were in, the ECGs read, and the patients had been prescribed either methadone or time-release morphine, he would hook them up to their drip stands and leave them to their journeys – farming out the 'prep' and 'integration' (the 'messy bits') to a psychotherapist in Stroud called Enzo. But, unlike most clinicians, John possessed the street-smarts of the ex-junkie. 'We need to ensure the addict's on clean supply – if they're not, we'll see it on an irregular ECG. These days there's so many impurities, and the dope's so *strong*, they even cut it with caffeine to keep people conscious. We see more and more horrible shit since Covid because of breaks in the supply. Like this black-market fentanyl they cook up in Mexico – the cartels are breaking that market – and they cut it with all sorts of nasty junk. It's made us nostalgic for honest-to-goodness old-fashioned smack.'

As for the sensitive issue of dosage, John follows his mentor Molly, who trained under ibogaine pioneer and ex-junkie Howard Lotsof. Their dosing protocols differ from most facilities. 'I did my first treatment in Baja. There's still tons of clinics there, like drive-throughs for opiate-tourists. They flood the junkie with single high-dose ibogaine, white-knuckle it, and then sling them out before they've fully come round. I'm not saying they don't perform a function, but for the aftercare you're on your Jack Jones. I learned the hard way, and it nearly fucking killed me.'

Instead of the flood dose, John was taught a different approach: gradually reducing the morphine as you increase the ibogaine. The process takes longer, eight or nine days, and the patient ends up receiving twice as much ibogaine overall, but 'it's less like pulling the rug from under their feet.' There's the temperament of his average client to consider. 'Opiate types don't tend to like having their head exploded by flooding them with psychedelics. They prefer to be hooked up in the corner and have a little nap.'

Given the incredibly long subjective experience (up to eighteen hours) and cardiac safety issues, it's not surprising that scientists are trying to tweak the molecule. Psychedelic Alpha reported on one group, led by Dr David Olson and his academic spin-out company Delix Therapeutics, who believe they have engineered an entirely non-hallucinogenic analogue of ibogaine, which they've dubbed Tabernanthalog.

As with all psychedelic drugs, the contemporary branding can be better understood if one knows a little history. French colonists brought back *iboga*, the bark of a local shrub (*Tabernanthe iboga*), from West Africa at the turn of the twentieth century. *Iboga* was used in various complex ways by indigenous tribes, notably as a ceremonial rite of passage for young men. As Josh Hardman explains, ibogaine, a naturally occurring alkaloid, was extracted from the bark and marketed as a 'stimulant' in France under the brand name Lambarene. At the same time, another pharmaceutical company took some of the same extract, mixed it with the belladonna plant, and marketed the compound as a 'tonic', under the name Iperton. Both products were sold on the basis of treating an unlikely array of overlapping psychological and medical conditions, including lethargy, muscle weakness, depression and infectious diseases.

Pharmaceutical companies have a history of adapting their products to the demands of the market rather than the dictates of science: the same drug, diazepam, that was used to treat a diagnosis of anxiety in the Sixties ('Mother's little helper'), was rebranded twenty years later to treat a diagnosis of depression (although you might rightly say that there's a conceptual overlap between the two diagnoses, which makes a single treatment less outlandish). In the case of ibogaine, we might want to explain the early-twentieth-century marketing as the exuberance of a snake oil salesman: corrupt undoubtedly, but lacking a basic understanding of either the mechanism of action of the drug itself or the 'science' underpinning the conditions it allegedly treats. Except for the fact, as Hardman points out, that today we are seeing similar levels of exuberance around the 'transdiagnostic efficacy' of many of the psychedelics. Psychedelic-assisted therapy appears to 'work' for depression and PTSD, but also smoking cessation and substance-use disorders, headaches, chronic pain, stroke rehabilitation . . . The list of possible applications gets ever longer, meaning that clinical potential has to be balanced against opportunists looking to ride the hype.

Being 'transdiagnostic' is currently in vogue, but for most of the last fifty years ibogaine has been marketed and used as a bespoke treatment for opiate addiction (though of course opiate addictions often bleed into other addictions, including alcohol and cocaine). Clean owes its existence to a separate tributary of ibogaine history. In 1962, the heroin addict and chemistry student Howard Lotsof accidentally discovered its anti-addictive effects when he took it along with five opiate-addicted friends. All of them experienced a subjective reduction of their craving and a reduction of the normal opiate withdrawal symptoms. Lotsof, the accidental first-person research scientist, went on to become a professional chemist and influential proponent of ibogaine as a treatment for addiction.

It had been decades since I last used what were conventionally classed as drugs or alcohol. Though interestingly, since I'd begun my psychedelic experiment, I'd inadvertently re-cultivated a chewing-gum habit which seemed to tap the water table of ancient craving:

one piece one day, forty the next, the familiar contractions as I approached the end of the pack. Why chewing gum? It was a combination of being permanently marooned in the oral stage of development and the aspartame sweetener. As the poet (and amateur neuroscientist) Don Paterson notes, 'the constellation of neurons that makes me a sugar addict will always be intact, and still clicks on like the Blackpool illuminations at the sight of a Twix.'[3]

The whole idea of me scoring an ibogaine 'experience' in the context of desperate opiate addicts felt wrong. On the other hand, Clean was wanting to diversify, broaden its horizons to include the leisure market in the form of 'psychospiritual' retreats. And how could I not remain curious about the kind of experience that 'the most powerful psychedelic known to man' generated? I had read research papers in which the hallucinations were described as high-resolution photorealistic visions with enhanced greyscale, or as a 'waking dream' that could include verbal interactions with 'ancestral and archetypal beings'.

There was also Daniel Pinchbeck's celebrated account of the *iboga* ceremony in his book about psychedelics and shamanism, *Breaking Open the Head*.[4] Working in traumatic brain injury at the time of reading it, I was shocked at his description of certain young tribesmen, taking *iboga* as a rite of passage, striking themselves on the skull with large sticks or heavy stones while undergoing their journeys. Shakespeare appeared to understand this psychology of double masochism: 'Where the greater malady is fixed, the lesser is scarce felt.'

Or in the CliffsNotes something like: *I won't feel like my brain is exploding if I really explode it with this big stone.*

Unlike me, but like the research scientists at Imperial, John seemed less curious about the experience. 'People talk about the experience – the insights, mysticism, ego death – but I'm all about neuroplasticity. It just depends on the person, I suppose. I've got two guys with me now: yesterday both of them were lying there like lambs for four hours while they dosed. I found out later one of them was reliving in real time his last eight reincarnations on earth: his face at different ages was coming at him on a giant revolving paddle and, as it smacked into his real face, a baby face would come out of the back of his head and then go round again, like the wheel of life. Meanwhile the guy in the next room on the same dose was bored out of his skull and kept asking for a cheese sandwich. For the most part you lie on a bed and let the ibogaine safely seep into your veins, like a blood transfusion in a hospital, only this is an Alpine-style cabin with ocean views.'

If Enzo thinks the patients will benefit from a therapy perspective, John will hit them with a snort of 5MeO, a form of DMT, of which more in Chapter 6.

'You mean the Toad?' I ask, using its colloquial name, as the drug is typically extracted from the glands of the Sonoran desert toad.

'No, I'm vegan. I use synthetic.' John explained how they used the well-documented 'cosmic spiritual bang' of the 5MeO to 'piggyback' on the ibogaine, no doubt accompanied by ash-smudging and world music. 'That's what we at Clean call our *psycho*-spiritual protocol.' It sounded like Charles Manson coming out of John's mouth.

'Bollocks, it's all *about* the experience,' said Enzo, 'psychospiritual consultant' slash therapist for Clean. 'Let me put it this way: if mushrooms are the Clangers, and acid is "Yellow Submarine", then 5Meo is *Star Trek* on steroids, and ibogaine and *iboga* are *Lord of the Rings* on . . .' He'd run out of drugs.

J. R. R. Tolkien. To me that meant dull, overlong and full of childish nonsense language. To Enzo it meant 'mythic', 'ancient', 'of the earth', 'brutally honest', but also – people in the psychedelic community are often inconsistent when it comes to anthropomorphising their favourite substances – 'paternal, sensitive, loving'.

Enzo was on a Zoom screen in his kitchen, somewhere outside of Stroud, as I continued to haver in the Pacific North-west. He was wearing a pink shirt unbuttoned to his chest, smoking a cigarette, taking some care to blow it in the other direction as he fed a two-year-old on his lap, who hid somewhere behind the same salesman-blue Italian eyes as her father.

Enzo was *all about* psychospiritual. Ibogaine is a fantasy quest:

'She' *called* you, the real journey was *inside* . . . The 'heart of dark-ness' was day six, when everything turned greyscale. 'You wake up in the bowels of Mordor, inside an orc's underpants, people swim-ming in all their shit, miserable, physically broken, wanting their money back. If we gave them a match they'd burn the place down, let them out, they'd run headlong off the nearest cliff. But if they can only hold on till morning, they wake up in glorious Techni-color, a new-born baby.'

Human or orc, he didn't say.

A folk theory of psychonauts had been taking shape in my mind that matched particular types of person to particular substances, their psychedelic preferences fed by the deep, half-understood cur-rents of temperament. It was a kind of psychedelic astrology. Applying the same theory to facilitators and therapists, Enzo's psychedelic 'star sign' would be ibogaine: epic, promotional, con-tradictory, and inescapably long-winded.

'There's a reason I'm in rural Gloucestershire, and it's not the reason I thought it was.' His stories began, without invitation, at a hundred miles per hour. 'I remember taking a stroll down Porto-bello one morning after an extra-long long weekend. Even the cars parked on the pavement looked drunk. I was totally fucked, to put it clinically. Standing outside this big white wedding cake of a house, I could see a man through the living-room window, an executive type, on his phone, making big executive gestures with his big exec-utive hands. Nearby, on the corner of All Saints Road the Jamaicans were jawing at the jerk chicken van. Opposite, the florist was giving chat to a girl, how the clematises were the same shade of blue as her eyes. I thought to myself, "None of it's real, it's all charades. Every-body's acting out the title of some stupid film without realising it, without even knowing the film's title . . ." I came home and told my wife: "I'm finished with this crap, let's move to the country." Within a month we were living in the middle of Gloucestershire. And guess what? It's the same pantomime here too . . . But that's not the real reason I came here. I'll tell you the real reason . . .'

Enzo broke off a moment to bark at his five-year-old off-screen, thereby spilling coffee down his pink shirt, scalding the toddler, who tipped her plate of eggs over the edge of the tray table in the commotion. More barking, and the extravagant cries of two young children. Was this Enzo's charade, I thought? Was the film he was unwittingly acting out *Hellraiser II* ? *Runaway Train*? *Everything Everywhere All at Once*? He'd already detailed his own long, drawn-out battle with addictions to heroin and crack cocaine, the multiple 'shameful, pointless' rehabs, culminating five years ago in his first psychedelic treatment. Now he explained how *iboga* had changed his relationship to all substances. 'I'm OK with a G&T with the wife of an evening, which is not the same as saying I crash in front of the TV with a crack pipe.' True: it wasn't the same.

'The real reason . . .' Enzo resumes his story. 'Look, I'm not proud of the way I *interact* with kids.' He occasionally seasons his street talk with therapeutic terminology, as though momentarily remembering the character he's supposed to be playing. 'My father was a terrible shouter. That was the intention I went into my first *iboga* ceremony with – to take a deep dive into unresolved anger, which is really unresolved sadness – but I'm getting ahead of myself.

'I was sitting in this circle in a tent somewhere in Sussex with a bunch of junkies. Within a few minutes of drinking the *iboga* this little voice starts up: *You fucking raud, you're so pathetic*—

'That's what She sounds like, and She continues like that, abusing me, calling out all my flaws, generally grinding my face in the filth. *Iboga*, I think, the voice of truth. *Why are you sitting outside the circle? It's because you think you're special isn't it? Why do you think this treatment's going to work? Nothing else has. You'll relapse for sure . . .*

'I think, She doesn't normally attack people in this way. Then I think, that's because I'm *such* a lowlife: She's making an exception for me . . .

'The attack went on for four, maybe five hours, subjected to the worst shaming you can imagine. I keep on resisting, wondering why She's persecuting me, until, out of the blue, the epiphany came: the *iboga* wasn't attacking me at all. She was *showing* me how I normally talk to myself, my inner voice that scores everything I do.

Only I'm so deeply inured to it I don't notice it's there, except when it leaps out to scald my wife or the kids. And as soon as I've joined the dots *iboga* says, *Thank God for that!*

'Then She takes me on this magical mystery tour round the world and into deepest space. *Do you want to meet God?* She says. *Do you want to go to the place that only a very few have ever been?* "I certainly do," I replied.

'Instantly I'm hurtling down a white wormhole, surrounded by beautiful bright light which gets brighter and brighter the deeper I go. I'm just as about to get there, to the end, to the bottom, to God – I can feel it in my bones – when there's this old-style movie wipe, and I'm in a field in Gloucestershire.

'It's early morning, the dawn chorus, mist on the ground, a few oak trees, etcetera. Otherwise, it's just a bog-standard field . . . Well, I was *so* fucking disappointed. I mean, talk about a comedown . . . And that's the pattern for the next ten hours: huge interstellar orgasmic build-ups followed by the same wipe to that same stupid fucking field . . .

'By the end I feel She's taught me a lesson about what scientists call the default mode network, and what I prefer to call "spending every waking moment with a super-chatty psychopath": my mind, in other words. It's a useful lesson, for sure. But really, I feel underwhelmed. Worse, I feel cheated . . .'

Enzo pauses for effect. The toddler on his lap plays with her eggs, blank-faced as though even at two years old she's already heard the story many times before. 'Three years later, a few months after I brought my wife and kids here from London, I'm driving back from my parents'. It's four in the morning. I look out the window and there's the fucking field, the trees, the birds, the mist – it's all there, exactly the same as the one She showed me. I get out of the car and jump over the fence. I'm standing in the middle of godforsaken nowhere crying my eyes out with how beautiful it all looks . . . That's God, right there. Her lesson as plain as the nose on my face: *Open your eyes. It's a gift, all of it . . . You never need use drugs again.*'

He lets it sink in, as though persuading himself one more time.

'What about the anger?'

'Still fucking terrible.'

Through it all I can't help thinking how we just can't resist *stories*. A trip experience of eighteen or more hours is made up of thousands, tens of thousands of individual episodes: thoughts, sensations, pictures, short sequences, snatches of conversation – all momentary fragments for the most part, like dreams. Yet all that gets smoothed, compressed, modelled, into a story – a trip report – as shapely and pointed as Enzo's. A 'lesson', really. The individual 'bits' of information are reduced to a kind of binary by our primitive narrative computer: story/not story. Our 'intention' beforehand and the process of 'integration' afterwards may be key to this process of construction. The intention acts as a prior that helps select – but also determine – what's relevant, a thread in the labyrinth that means we will find the only way out. 'Integration' is a password for that part of the reality we are prepared to digest; the story we will keep on telling ourselves and then others, until it's as smooth and well-worn as an old man's shins, until all that reality has been transformed into a turning point or anecdote, a homily of sorts. Or, as here, a paragraph or two in a memoir; the story of a life rather than life itself.

Addiction is its own brutal form of story-making: the reduction of everything possible to one thing alone. Enzo's story was that of an addict who becomes a therapist, a therapist who is really a salesman: basically, a story-teller with a financial incentive. At some level, he knew that I knew this, and he wanted me to know that he knew that I knew too. 'What happens to a narcissistic sociopathic addict after he's been treated with *iboga*? He opens an ibogaine retreat centre.' He had as few illusions about the business he was in as he did about himself, and that knowingness had woven itself into the fabric of his pitch. 'Sad to say, there's an awful lot of people in the field who look at addicts as cash cows; they're so fucking desperate they'll pay twenty thousand US per week for a hundred and fifty dollars' worth of ibogaine.'

Psychedelically assisted addiction retreats, a tributary of the international wellness industry, are set to be worth $1.2 trillion within the next five years. I'd looked at these clinics online, in Mexico, Portugal, Costa Rica and Switzerland: images of palm trees, frisbee-tossing, water sports and chia-seed smoothies, rather than filthy needles, rotting teeth and deep-vein thrombosis. The reality of addiction for most is very different, and has nothing to do with wellness retreats. Most of the homeless addicts I worked with in Leeds had never seen the sea. And the only retreat centre the vast majority of addicts will ever see is a prison.

Addiction does not distribute randomly in the population. Almost without exception, the street addicts I worked with in Yorkshire came from deprived socio-economic backgrounds. According to the Drug Policy Alliance, nearly 80 per cent of people in federal prison and almost 60 per cent of people in state prison for drug offences are Black or Latino. In the US, because of laws that remove the right to vote for felony convictions, one out of every thirteen Black people of voting age has had their democratic rights removed. Which is to say that for many, addiction is a complex social problem before it is a medical one. Addressing the problem of addiction via treatment of the individual – whether via psychological or chemical means – is already an inadequately narrow approach.

In the 1990s, when I was working in the field, there were expensive, exotic medical treatments available for addiction, like the week-long anaesthesia in a clinical setting to bypass withdrawals. On a few occasions families were able to raise enough funds to send their son or daughter, and they learned the hard way: whatever the efficacy of the treatment, the overall situation remained unchanged. It would be a few weeks, perhaps a few months, before they were back on the streets, more humiliated, more impoverished than ever by the latest expensive failure.

'I believe loving money is one of the greatest gifts you can give humanity at this time,' said Azrya Bequer, one half of a husband-and-wife duo behind the psychedelic retreat start-up Beqoming, which I read about in *UnHerd*.[5] Beqoming aimed 'to serve the 0.1 per cent of the wealthiest people in the world to start making better decisions for our planet and our future generations' – through facilitated ayahuasca ceremonies . . .

The idea of a deluxe, individualised psychospiritual retreat with a substance so heavily associated with addiction treatments, which itself was so heavily associated with poverty, continued to pall. My own consumption apart, weren't they just trying to elaborate the applications of a drug to broaden its market share?

I shared my concerns with Enzo, who looked uncharacteristically thoughtful for a moment. Then he told me to speak to Molly, John's mentor, who, he said, ran a centre in Mexico with a much more developed psychospiritual 'protocol'. 'Molly trained under Lotsof. Her treatments are about as authentic as you'll find.'

I turned the car round a second time and drove south on I-5 once again.

Molly's face is a little too near the screen. Up close she looks as though she's been in a fight. The night before, a young Israeli man had turned up at the door of her the clinic demanding a treatment. 'He was pleading with us desperately. He'd come straight from Lima after thirty consecutive nights of ayahuasca. His shaman told him he needed something *stronger*.'

Not your averagely curious psychospiritual punter, then.

Molly already has a waiting list of several months. 'There's more unreal expectation with ibogaine than any other substance. I blame the hype: all these people believing it's years of therapy in one night, that they're either gonna get a miracle cure or, if they're in it for the *experience*, that they're gonna have the most intense fireworks. "Prepare for disappointment," I tell them. Ibogaine is an intelligent plant; it rides the waves of expectations and dismantles them one by one.'

Most of the addicts she saw, though desperate, weren't, in her view, deeply motivated to change. They still had in their minds a 'medical model' for treatment: a patient receives another drug, only this one makes him better. But ibogaine was different: 'It needs you to make a real investment before it shows you anything.'

Molly went slowly with the treatments: low doses over days, alongside therapeutic work on the patient's motivation. 'You can't just break open the head like Pinchback says – you have to crack it gently, a little at a time. If you're lucky the chick hatches.'

Like Enzo and John, Molly was an ex-junkie. When she went through her own treatment with ibogaine in 2005 only 3,000 people were known to have used it. The figure is growing at a rate that disturbs her. 'We don't have the governance and structure to handle the demand safely. That's why there's all these pop-up flood clinics in the north of Mexico. Officially there are twelve deaths a year from cardiac arrests, but we hear of many more than that which get brushed under the carpet. That's not the only thing. The flood might get you clean for a bit, but it's fragile. The effects are likely not to last.'

The increase in demand also meant that non-specialists were opening up provider centres across the country, like the dentist who ran ibogaine treatment alongside root-canal surgeries in a one-stop shop. With growth came interest from the narcos, who could in the years to come, she thought, view the territory as their own. Molly knew one ibogaine provider who had already been threatened with having her eyeballs scooped out if she didn't treat a trafficker's son. The insane demand for addiction treatments had meant Molly had had to abandon her own 'psychospiritual' track for the time being.

On cue, there's a loud banging on her door, followed by shouting in Israeli English. She has to leave the meeting immediately. I'm left staring at a blank screen. To have an ibogaine experience in Mexico would mean either stepping over an addict in desperate need, or risking the chicken run of a pop-up clinic. Neither sounded like a good idea.

This particular psychedelic experience was proving very difficult to score. I was continuing to drive round in ever narrowing circles in Oregon, my road trip's K-hole, when I received a call from Carlton Bates, who'd been given my number by Enzo. Carlton was CEO of TrueYou: not the Pynchonian brainwash programme it sounded like, but a start-up offering ibogaine retreats in the Caribbean, though Carlton himself was still based in Wakefield.

'Childhood conditioning and epigenetics have meant that many of us have bad coding that can affect our operating system and software.' Carlton was fluent in contemporary neuro-computational psychedelic speak (*#psychedelic-ese*). 'Ibogaine is a defragmentation, a restoration back to factory settings.' But my conversation with Jaron Lanier had led me to see the technologising of psychedelic language had more sinister overtones. Besides, the technical metaphors don't sound quite right in Carlton's mouth: he looks more bohemian, more anaemic *fin de siècle* poet than plant-based tech exec; someone who just about knows how to turn a computer on is my sense of him. 'It doesn't get rid of the bad coding, but it gives you the opportunity to put Norton antiviral software in. In my case, I could feel myself losing certain files which were not doing me any good, I literally watched them flying away. I could feel space being freed up, a space that allowed me to re-boot my life.'

The crux of the sales pitch was the opportunity to be True-You's 'first psychospiritual client . . . our first client, in fact', in exchange for a profile in *Esquire*. Enzo and John, both old friends, would be joining the team, relocating to the Caribbean in due course; Molly would offer remote consultations. In other words, all the people I had spoken to – from Mexico, British Columbia, Stroud and, in Carlton's case, Wakefield – about a derivative of a West African plant that was being used to treat the North American opiate problem in countries other than North America (where it was illegal) – were gathering together at a seven-star resort on a Caribbean island of a former British colony. It sounded like the synopsis of some berserk psychedelic novel. Carlton explained that much of the Caribbean offered English law and order, alongside low-cost labour, sunny beaches (and levels of corruption and nepotism that, for all of its promise, England could still only dream about). It was economically appealing to establish a foothold there: seven beds to begin with, a two-week stay costing £25K, and crucially it was legal – though fragilely so – with the right kind of licence. 'Thankfully our chief psychiatrist was best friends with the Prime Minister when they were at school.' Though his talk was smooth and boosterish, Carlton's Zoom face was baleful, as though his software had been corrupted by a virus of unspoken guilt.

'What do you say: an ibogaine retreat by a crystal-clear turquoise sea?' His heart just wasn't in it.

'I say thanks, but no thanks.'

And I told him all my reservations: about the model of addiction treatment; about the mercenary-sounding attempt to capture the psychospiritual market; about what I knew of the psychedelic retreat industry in general.

To my surprise, Carlton nodded in agreement, looking sadder and younger as the moments passed. 'You're right. But still the fact remains that this is an extraordinarily powerful plant.' The recklessness and profiteering had not completely contaminated his idealism. 'I believe that it's possible to do so much more than detox like the flash clinics. That, while it's by no means guaranteed, the medicine, alongside the right kind of pre- and post-service, can fundamentally change the direction of people's lives.'

Carlton opened up easily, like a patient ready to discharge his burden. He'd always felt deeply drawn to Africa. He'd lived there, worked in various mining concerns trying to improve the conditions of the labour force, had experience negotiating with lawyers and civil servants, in writing policy, in dealing with corrupt and hostile governments. He'd also encountered the miracle of *iboga* first-hand, spending months in a Gabonese village undergoing initiation with the Bwiti tribe. 'I'm not going to try and persuade you otherwise. I can't. All I can say is, I'm going to try and steer things in a way that's true to my original vision. But I'm also realistic; there's a large board, complex political dealings, the dictates of investor returns, etcetera. It's more than likely I'll be edited out by system requirements after a year or two. I'm just not your smooth, face-making CEO.'

He sounded wistful. He also sounded *authentic*.

If the whole shtick was a grift tailored to me, then the next thing Carlton told me was its coup de grace.

Before he turned corporate, Carlton had done a PhD on Merleau-Ponty.

A phenomenologist, at the heart of the international ibogaine retreat industry!

It was strangely reassuring. First Carlton had tried psychedelics, then he'd gone onto the hard stuff: continental philosophy. 'Merleau-Ponty is behind us historically, but really he is ahead of us. We still haven't caught up with him.'

We agreed that the French philosopher had been almost completely neglected by neuroscience, even psychedelic neuroscience, but that Merleau-Ponty's experience with pharmaceutical mescaline had allowed him to suspend the analytic attitude that atomises the world, and return to a 'wild' perception that was more authentic, more real. Carlton extended this idea to clinical treatment: ibogaine was not about patients receiving medicine from a drip in their arm, but, as with the richest philosophy, a creative opening up that allows different senses and perceptions to 'inter-communicate', to 'hyper-associate', 'see the world as it is in its original character'.

Just another story, another type of shtick, maybe, but my kind of shtick. I turned the car around again.

The Caribbean. A seven-star resort. Cold-brew ibogaine by the pool. But first of all, Paddington in the rain, waiting for the delayed Stroud service. I was back in England for my *'pre*-pre-treatment'.

Enzo met me in a battered Volvo hatchback decorated with foam coffee cups, Red Bull cans, vape capsules and cigarette butts; a different kind of well-being mobile for a different kind of psycho-spiritual therapist. We drove through farmland to Frith Wood, where young couples with buggies and older people with dogs were setting off on Sunday walks.

'Think of your journey as a play in three acts: a walk in the English countryside on mushrooms; a flood dose of ibogaine in the Caribbean; then a shot of 5MeO on Day Seven to really blow the doors off.'

Apparently everything was ready for me over there. Enzo would head out with me the following week; his wife and kids would follow in a couple of months. Despite the field where God was found, he didn't want to stay in rural Gloucestershire with a young family. If nothing else, he'd learned that life was about openness to new possibilities. And one possibility in particular. He paused and turned to face me: mid-fifties, aristocratic beginnings, then decades of grifting and struggle. 'This is my last shot at generational wealth.'

He handed me a pinch of dried liberty cap mushrooms. 'It's time to explore your intentions for the ibogaine journey.' A low-dose nature walk amid ancient English beeches. 'Don't force it. See what comes up.'

I cleared my throat, about to 'explore' my inner realms, when Enzo broke in. 'Can I level with you?' Just like the girl in the documentary.

Actually the retreat centre itself *wouldn't* quite be ready in time to host me. There were problems with the builders. They'd arrange a nice Airbnb by the beach instead. There'd also been a little political chicanery, which had meant procuring the necessary licensing had been a little trickier than expected. '*Should* all be in place by next week.'

We walked in silence for a while.

'Look, mate, can I really level with you?' There were already quite a lot of levels in play, and the mushrooms made it even harder to keep them level. 'This whole TrueYou thing is so stressful . . . When I was using I would cheat girlfriends, rip off mates, steal from parents and still sleep like a baby. But getting involved with this company has given me more sleepless nights than you could ever imagine.'

This was *his* pre-treatment, not mine. I wouldn't be going to the Caribbean to take the most powerful entheogen known to man, not for all the frisbees in California. We arrived at a quiet glade in the midst of the ancient forest. I'd been to Frith once before, on a long cycle ride with Danny a few months after he'd been diagnosed with ALS – proof, he thought, that his late-forty-something fitness would extend the years he had left past the normal clinical range. He was dead within a year. I wondered what Danny, a deeply sane clinician who took an unusually rich, anthropological interest in the lives of his patients, would have made of a seven-star treatment centre for addicts. But then there's what desperation can do, from which nobody, in my experience, is entirely immune. In the weeks immediately after he'd lost the ability to swallow his soup or pedal up the hill by the hospital, Danny had seriously considered flying to China for a 'revolutionary' stem-cell therapy he'd seen advertised, when a few months before he'd have recognised it instantly for the heinous, evidence-less exploitation it was.

Generational wealth: Danny had left a generation of five children without a father to support them. Meanwhile Enzo was picturing how, post-ibogaine, patient and therapist would toss the old self like a frisbee on the beach.

'Stop talking – just, *stop!*' I turned somewhat dramatically to look Enzo in the eye. 'I want you to remember what I'm about to tell you, Enzo.'

Maybe it was the mushrooms, but the whole forest appeared to fall silent.

'One day before too long you're going to find yourself looking across your desk at a new client: someone whose life has been utterly torn to pieces, and you'll know full well that your treatment isn't going to work for them. It might give them the reprieve of a week or two, but they'll be back to their old ways – only now, on top of everything, they'll be ruined financially, because they put every last egg in your basket.'

Enzo was pulling hard on his cigarette.

'You'll have all these pressures; from the board, from your wife, from your screaming kids, from your own unfulfilled dreams. And in that same moment you'll have the chance to feel what it's really like to be *authentic*.' I could feel my whole body hardening with anger. 'You're going to look that wretched boy or girl straight in the eye, sitting alongside their desperate mothers and fathers, and you're going to tell them the truth. I know you'll do it, Enzo. I know you'll do it, because you'll remember this moment, this forest, these mushrooms, the pressure, this speech, the look in my eyes, all drilling into your hippocampus . . .'

Brainwashing, *brain-cleaning*, trying to reach the TrueYou . . . It was a kind of therapeutic abuse in reverse; the patient exploiting the suggestibility of the mushrooms for his own ends. I didn't care. That was my 'intention'.

Now I had to put up with the fact that one of my trips had gone missing, leaving me marooned in London, back where I began, 5,000 miles east of the sacred psychedelic ley line.

An underground contact put me in touch with an *iboga* practitioner on Dartmoor – one last shot at the orc's underpants. Ben, an artist and landscapist, had studied with other initiates in a Gabonese village for two years before serving *iboga* to people he'd discerned had 'a calling' for it. Then he'd stopped doing it.

The difference between *iboga* and ibogaine, between Ben and Enzo or John or Molly, between healing ceremony and addiction treatment, could not have been starker. No bluster, no promises, no computer metaphors, no frisbees: Ben's speech was steeped in indigenous culture and ethnobotanical knowledge. Within a few minutes I actually learned things other than how to duck a sales pitch. *Iboga* could be served as sawdust with a drink, or as a fermented liquid, in the form of a plantain 'bleeding *iboga*', or even as an enema, which divides opinion in a strongly patriarchal culture. Also, that it's hard to transplant the tree to other locations: it grows as a sub-storey shrub, surrounded by giant trees, so successful growth depends on a complex interrelation between different species. The 'immigrant' can find it hard to make stable relations elsewhere. Often the tribe is shown the shrub's location by elephants who feed on its fruit, creating direct pathways to it. This has led to an unfortunate 'natural' synergy between harvesting *iboga* and the ivory trade.

'Is the experience *Tolkienesque*?' I ask. It was the kind of thing the 'ibogaine crowd' might say.

'It could be, I suppose, if you're a Tolkien fan, but not inherently.'

Ben was frustrated by the conflation of the two 'wildly different' scenes. Though ibogaine had originally been extracted from the *iboga* plant, these days it was almost always a synthetic analogue. Josh Hardman had told me that there are a growing number of for-profit biotechs focusing almost entirely on analogues, derivatives and tweaked versions of ibogaine. These might offer better safety and efficacy profiles compared to natural ibogaine or *iboga*, but the prospect of new IP and patent protection around such molecules 'certainly sweetens the pot'.

'Would you call it a *de-fragmentation technology*?'

'I would not.' Ben laughed sardonically. 'But then I'm a gardener. My metaphors are greener, more organic, but they're still just metaphors.'

The psychoactive material comes from the root of the shrub; 'root' from *radix*, the source of 'radical', Ben explains. Thus the material works not in the astral planes, not on magic carpets, or the Starship *Enterprise*, but by its *humility*, its capacity to ground people, 'to help them walk with roots on the earth'.

'What makes it effective in treating addiction?'

Ben told me that, compelling as the evidence may be, it was provisional and circumstantial. To his mind it was an unfortunate, medicalising, Western association. Though there was increasing Tramadol dependence in West Africa, he explained, and more long-standing alcohol abuse, there was no history of opiate addiction. Rather, *iboga* is used for many things because of its spiritual qualities, which might include psychological problems but could not be reduced to them. There was a functionalism to plant-based medicine even in West Africa, but *iboga* was exceptional in this regard. 'If you're walking through the jungle with a Bwiti priest, he might point out the particular, specific properties of each plant. "This one heals X." "That one heals Y." But when he comes to the *iboga* shrub he likely leaves the particular behind: "This one *saves*," is what he says. Meaning, it connects you with a realm of consciousness, out of which all the cares of the world flow.'

Ben explained that *iboga* is unlike most medicine, unlike ibogaine even, because its action depends on disrupting safety, on breaking any kind of container. Risk is necessary to the experience. Sometimes it's the wrong kind of risk: he had witnessed Western medics saving the life of a man in cardiac arrest whose condition the Bwiti leaders had simply overlooked. 'But the risk is not just in the plant. It's in the person's mind – it's demanded of them. The plant asks, "What are you bringing to the table?"'

It occurred to me – a little superstitiously, perhaps – that this was the real reason I had got nowhere when it came to curating an ibogaine experience. I was a whimsical experience-tourist, risking little. At least, that's what I thought at this point on my journey.

'Whatever the science might say about neuroplastic timelines and maximising critical windows, *iboga* demands a different kind of relationship all together. The Bwiti believe . . .'

And before I tell you what Ben thinks they believe, I am compelled to say that, as earnest and reliable as Ben sounded, it was a well-spoken, white painter-gardener from Kent who was doing the speaking and not an indigenous person, and therein lies a vast and multidimensional problem that frames all psychedelic usage, of which more later.

'. . . The Bwiti believe that once you've ingested it, it's in you forever. It will unfold and grow according to the nutrition you provide for it. Which all makes it sound so organic, and natural, and wholesome, when really it's the most unimaginably divine chaos, that one must learn to adore . . . Like the head-cracking music the pygmies make during the ceremony, you slowly yield to its unlikely beauty: big bang after big bang, it just keeps on going. You have to develop the capacity to live with it . . . Or close down.'

'Is that why you stopped? You couldn't keep on living with it?'

'I certainly needed to create space for the experiences I'd already had. Painting and gardening are perfect for that. Integration's not a couple of sessions with a therapist; it's not about fixing or incorporating the meaning of the trip, because the meaning keeps coming at you, changing with you . . .'

'The story has no end.'

'Exactly. That, and the fact I was running ceremonies for others with only two assistants. In Africa, running a ceremony is the work of a whole village. But I adored it. I miss it.'

'Any chance you'd come out of retirement, for one last ride?'

'A few months ago I was offered fifty thousand pounds from the son of an Uzbek billionaire. I used to charge four hundred quid for three days of non-stop work. I still said no.'

That was the end of our conversation. But it was also a beginning, opening up a completely different way of thinking about the relationship with psychedelic experience, a way fundamentally at odds with the psychedelic retreat model I had encountered thus far, that so readily piggy-backs on all those medico-therapeutic findings.

As for ibogaine, the experience of attempting and failing to curate a retreat was provocative and frustrating in equal measure. At a banal level it reminded me of the 'good' old days, how tricky it could be to score: I mean, logistically speaking, just how difficult can 'drug-dealing' be? In retrospect I'm grateful for all those times the arrangements fell through – it saved me thousands of pounds while thousands of aggregated hours were saved for those forced to listen to me. More seriously, it provided a window on the different ways the commercialisation of psychedelic medicine might play out; on the strange symmetry between the intractable chaos of the patient cohort and the providers; the degree to which finance affects set and settings as much as clinical judgement; the potential compromises in the safety and reliability of the treatment for what were already vulnerable consumers. It also made clear how such treatments are reinforcing the concept of addiction as an individual disease, rather than a complex social, economic and cultural phenomenon.

Can I level with you?

It sounds a little New Age-y, I know, but there seemed to be a strange principle of substitution at work under the seeming chaos of this current escapade; of virtual reality for psychedelic reality, and of both for reality itself; of *iboga* for ibogaine; of an asexual African shrub for 'She'; of Canada and Mexico for America's drug problem; of a general 'tonic' for a specific addiction treatment; of one drug (ibogaine) for another (heroin); of a synthetic analogue for a tree bark; of Africa for the West; of an addiction remedy for spiritual healing; not to mention of seven stars for a one-star drive-thru; of imagined luxury for real-life desperation; of one therapist for another, and another, and another; of Stroud for the Caribbean and the Caribbean for nowhere and all of it for tribal West Africa. One substitution after the next, as though I'd navigated the whole trip by the position of Venus, thinking it was the North Star. But somehow this accidental course offered its own coherence, returning me, as it were, to where I started, the same conviction: the actual experience was missing.

The problem with ibogaine, as opposed to *iboga*, indeed with any of the synthetic analogues being tweaked and honed to produce ever more specific effects for supposedly specific diagnostic conditions, was really identical to the dark prospect of technology, as described by Jaron Lanier: both were a way of 'instrumentalising' our experiences. Increasing the 'specifications' – either the reality of the simulation devices, or reducing the side-effects of the drugs – was in keeping with the humanist drive to improve, and reduce inefficiency. It was all of a piece with progressive determinism, and the international 'retreat' market rode on the back of it.

But it didn't have to be like this. It didn't have to be 'therapeutic' in this narrow, functional, reductive way. Before I left Lanier's home in Santa Cruz, he showed me a piece of VR software he'd written forty years before, one of his first efforts, driven by 'skills-training' functionality but growing out of his obsession with exotic musical instruments. On the computer screen was the low-resolution, craterous surface of a planet scrolling on its axis. There were objects. Wearing a special 'haptic' glove, one could point to the objects and fly towards them. As the objects came into view it was clear that each was a different musical instrument, gaudily coloured like a cartoon. But though familiar, the instruments were novel, strange amalgams, centaurs – half saxophone, half guitar, a piano with the body of a trumpet. The glove enabled you to pick up the instrument and, by trial and error, find a way to make it play, by manipulating strings, valves, buttons, keys. In this way the mechanics of each instrument, its 'idea', were understood, and one could 'practise', learn to make better note sequences or chords. Once sounded, the notes continued to loop around planetary space as one flew off in search of the next instrument. As more instruments were played, the sounds combined and created a bizarre symphony of centaurs, a world cacophony by a drunk Stockhausen. The planet's lurid colours and strange shapes seemed to leap more furiously into life.

This was a piece of low-tech psychedelic art that caught something of music's reality, engineered as a skills-training instrument by someone profoundly uninterested in the drugs. Because instruments could be more than instrumental, art could exceed the limits of 'functionality', and experiences could be more therapeutic than any programme or protocol we might seek to constrain it by.

Now, can I really level with you?

Because even without ingesting a psychedelic drug, the trip made me think seriously about my own purpose in attempting to acquire these experiences. What, if anything, was I risking, to use Ben's formulation, over and above any kind of touristic, bucket-list acquisitiveness? What, if any, was the relation of risk, or its absence, to my recurrent anxiety in the face of drug-taking? For the plants to do the work something genuine and deep needed to be put on the line; the richness and value of psychedelic experiences depended on the radical loss of control, on uncertainty, on unpredictability of individual responses, and developing a long-term relationship with these kinds of states. It was a practice, a discipline, in the manner of a religious practice. Something would be lost if this was subjugated by design improvements, or by removing them wholesale from their native contexts to a Caribbean beach. And, as Lanier and William McCaskill and others kept wondering aloud, the improvements and subjugations kept outstripping our moral capacity, which lagged behind. It meant, paradoxically, that, for all the science, we really had no idea where we were going, or whether we could cope with it. The future was bound to be a psychedelic trip whether we swallowed it or not.

It seemed the right time to step outside the medico-therapeutic model altogether, and into a spiritual/religious one; because, to paraphrase Ken Kesey, sometimes the need for a mystery is greater than the need for an answer.

PART TWO

5

On Harmony

Once again I am encapsulated with 500 others in an aluminium lozenge, soon to be hurtling through the skies at ungodly speed. As the flight attendants do their air-safety charades, all I can think of is my friend Will's trip report. It was his bachelor party weekend. Breakfast-drunk with other stags in the Heathrow boarding queue for a flight to Riga, a seditious-looking best man hands Will a small piece of blotting paper, printed with comic art. 'Say goodbye to single life.'

Despite the groom's sputtered protestations, the rule of the weekend was 'complete rhinectomy' (the party were medics for the most part), meaning 'no nos' (nose). The wedding would take place the following weekend at St Nicholas's, a picture-perfect village church near Leamington Spa. The current jaunt, three nights of Baltic delinquency, was conceived as that English idyll's dark mirror. It was – unbelievably – Will's first time tripping. It was also – more unbelievably – his first time flying. The prospect of either petrified him. Together they were unimaginable. One of the ushers reassured him with signature *Schadenfreude*: the chances of the aeroplane exploding because of a hidden bomb was one in sixteen million, which rose to one in 243 million if there were two bombs on board. This, he explained, was why many people choose to carry single bombs with them when they fly these days.

The hallucination of normality that allows most of us to forget we are about to travel 30,000 feet off the ground at 600 miles per hour had been punctured by the effects of the acid, so different permutations of envisaged catastrophe were all Will could think of. His set – as a doctor his specialties were in emergency medicine and critical care – meant that his associations ranged over horrifically realist renderings of death which, under the influence of LSD, took on a deeper, supernatural horror, exceeding the conventions of physics, anatomy, biology. Added to that, the newly inserted thought that everyone on board had explosives strapped to their abdomens, even the screaming babies – screaming because they knew what was coming . . . 'Say goodbye to life.'

Only a species of visionary, psychedelic faith saved him: the flight attendants, signalling airplane orientation, miming emergency protocols, now flattered, by the groom's unblinking attention, to have an audience for the first time since the late Sixties, became priestesses, their gold- and purple-trimmed two-pieces now tailored chasubles and stoles, presiding over a high feast. Though these immaculately made-up Slavonic women in no way resembled the puffy-faced late-middle-aged bachelor priests from his choirboy youth, their mime and semaphore had been transformed into the charade of Eucharist. It became increasingly apparent to Will that it was only their holy mime, drawing down Grace from on high, that kept the plane aloft. With a priestess stationed between the rows of seats every twenty metres, the vision regressed endlessly, and he was transported back down the church aisle to the choir stall of his childhood, singing the *Nunc Dimittis* and *Magnificat* at evensong in St Nicholas's, the same church he would be married in next week. *Death by crucifixion. Resurrection by oxygen mask.* The hostesses handed out the host in the form of small packets of cream cheese pretzels, the blood as criminally expensive Cabernet Sauvignon miniatures. The religious faith that had lapsed long ago during those painfully tedious Sundays was born again with a fanatic's zeal.

Though the faith itself did not last beyond the nine hours of the trip, the memory of it did, and thus this traumatically mystical experience (trauma and mystery as one, not the discreet categories suggested by the literature) gave a whole new dimension to the 'I do' Will pronounced the following week, looking into his new wife's eyes before the altar of St Nicholas's, in the sight of God and in the face of a holy congregation, with a depth of sincerity previously unavailable to him.

It also led to a second pledge, a silent 'I will never', underneath the 'I do'. Though he loved her with all his heart, and they undoubtedly had many shared passions to look forward to, the recreational use of psychedelics would join the ranks of Zumba and book club as something Will's new wife must do alone.

By the time the attendants on my flight are strapping themselves in for landing, it has occurred to me that my doctor friend's experience was also an allegory for the marriage of two ostensibly ill-suited psychedelic models. Medicine drops acid and calls on religion to save its life.

The minister, his long hair braided in a single serpentine plait, is seated half-lotus behind a low altar. The altar's centrepiece is an arrangement of murky plastic bottles, each containing a different decoction of the 'sacrament', around which are assorted crystals, *mapachos* (Amazonian cigars), a block of the aromatic 'healing' wood Palo Santo, hawk feathers and some breath-freshening mints. By the side, there's a leaning tower of stacked buckets for puking. The minister's brilliant white poncho with peacock blue trim has something of the cut of a priest's white chasuble on a saint's day; a saint like the Franciscan mendicant Junípero Serra, who came by this way 300 years before, intent on changing hearts and minds by word or miracle or sword. Dedicated, as this minister is, to the foundation of a holy church.

The Church of Sonqo (*Sonqo* from the Quechua for 'heart'), of which Oberon, our minister (on Aurora's recommendation – 'the dude serves *strong tea*'), is the current executive director, has a guest list of 1,500 nationwide, and four full-time ministers running ceremonies most weekends in seven different US states in which the visionary Amazonian plant brew ayahuasca is served. As ayahuasca remains a 'controlled substance' here, the ceremonies are convened via invitations to guests and vetted friends. Currently the church ships its 'sacrament' from Peru. A few years ago it bought land in

Hawaii, which shares the same tropical climate as the Amazon, allowing for the ayahuasca vine and its partner plant (in Sonqo's case chacruna) to grow prodigiously. It takes seven years to establish a viable harvest. It is hoped that within the next couple of years Sonqo will be able to serve 'home-made' American ayahuasca.

The church has been meticulously constructed. Oberon (not his real name, which is much stranger) and his fellow ministers operate under the guidance of a metaphysical creed ('the Ten Commitments') and a physical board of directors, including healers and therapists, but also a financial analyst and a lawyer to guide more hard-nosed decisions on future directions. As well as weekly retreats, Sonqo runs an online Sunday service, outreach programmes, equality and diversity initiatives, integration workshops and scholarships for the economically challenged, while also participating in several different indigenous reparation schemes.

Sonqo's thoughtful construction reflects its members' dedication to a spirituality based on a mestizo adaptation of an ancient Peruvian lineage which has in turn been adapted to a North American context. It also reflects a legal strategy: the church gives its ministers and members some recourse to legal defence in the event of prosecution. In fact, it is so well designed and 'scalable' that a celebrity patron slash psychedelic apologist has approached the church with the offer of funds to actively sue the Drug Enforcement Administration, rather than wait for prosecution, on the grounds of religious persecution. If it decides to proceed and is granted exemption under the Religious Freedom Restoration Act (as the Native American Church and its peyote rite have been), then Oberon and his team would be able to surface from the underground without threat, providing a blueprint for the religious use of all psychedelic medicines in the US.

A solitary candle is the only point of light in the pitch-black temple, where more than forty congregants line the perimeter, buckets and folds of toilet paper at their feet, dressed in coutured white like hip ghosts. The candlelight casts Oberon's long figure onto the wall, folding across the join of the ceiling where giant Giacometti-thin arms consecrate the sacrament below with an aspergillum of olympia, mint and mistletoe, under the whisper of Spanish and Quechua incantations. He lugs deeply, un-hieratically, on a *mapacho* and, cheeks puckered, blows thick plumes of creamy smoke over the medicine like the figure of wind on an ancient map. 'Tonight, dear friends, we have an opportunity to dive into unknown, unexpected terrain, dedicated in the pursuit of harmony and coherence between all beings seen and unseen.'

Towards the front of the temple sit the meek, who along with the rest of us will be tended by two 'ceremonial assistants': a young woman with her dying dog, a blind man with his guitar and drums, an older lady on crutches. Further back, the well-heeled majority: two ER doctors, a group of Web 3.0 developers, polyamorous Insta influencers, assorted hipsters in their thirties whose Sprinters with boutique hotel-level interiors are parked on the forecourt of this remote New Age retreat centre. Somewhere among them sit Martin and Camilla, two friends in late middle age I've invited from North London. Tonight is the third consecutive night of 'sitting', my friends transformed from 'drug-naïve' retirees to puke-happy, shell-shocked veterans in a little less than forty-eight hours.

'For many of us it has been a trying, exhausting time: it's taken a lot of resources just to show up today . . .'

There have been moments of rare beauty on the first two nights, not least the visual extravaganza of 'eyes-closed' hallucinations whose strangeness is only exceeded by their 'realness'. Recent neuro-imaging studies indicate that ayahuasca enhances the intensity of internal imagery to the same acuity as normal vision, even if the new 'normal' means detailed rendition of a man being eaten by a piano in the shape of a crow. (I could even discern the minuscule, anatomically accurate hook on the end of his obsidian bill.) But the main themes of my first two nights were not beauty, however surreal, but fear, confusion, endurance.

'I believe there is great value in this practice even when we can't consciously detect what that is. Undoubtedly there is a lot going on that is unconscious, at the level of brute physiology. The plant is famously plastic in its effects. The movie that we get not just to see but to participate in is just one facet of the whole picture, a picture that is by its nature unfolding, beyond the provenance of any author.'

Oberon talks with the fluid prolixity and precision of ChatGPT as it might be a few years from now, 'musically' if that music were Bach counterpoint, the melody elaborating in complex, novel ways that make cliché an impossibility, and yet somehow managing to smuggle in human warmth under its crystalline form. He also sounds straight, preternaturally straight for a professional purveyor of head-bending tea. 'As we prepare for the journey tonight, I pray that we might have some insight, we may gain even a modicum of clarity as to the interface between those other worlds and the world we conventionally inhabit, which might shake up or otherwise unsettle our regular thought-grooves, the sense of who we are, of who we might become, in ways that remain rich in their suggestiveness long after we dutifully return to the responsibilities and roles that await us tomorrow.'

In a few introductory sentences Oberon had just mapped out a decade's worth of medical research: unconscious processing, insights and challenging experience, the disruption of 'thought grooves' or 'canals' in the default mode network, post-ceremonial integration and the ongoing plasticity of the so-called 'aya-glow'.

Even though it's dark in here, the diffused moonlight allows me to make out the willowy shadow of Joshua, a Bitcoin retiree in his mid-twenties who travels everywhere, like an updated Dylan song, in a rented Tesla with his Applehead Siamese cat. For Joshua the world has become an endless pilgrimage to the next ceremony, his only remaining ambition, as he tells me in drawl so gentle and lacking in ambition it can barely make it to the end of its sentence, to reach *saturation point*, where man and plant become one. Opposite him I make out the singularly upright form of Zak Talon, rapper, aikido master, spiritual entrepreneur and creator of Talon's Lair, the world's first Tantric theme park slash gym. On the introductory round robin Talon introduced himself as '—', meaning he didn't 'do' names, that no name could mean anything in the face of Talon's Lair and Full Tantric Union. Coming down from my own half-baked version of ego-death (*ego-quite-ill*) on the second night, I resolved to substitute '—' for 'I' in all future conversation as a way of guaranteeing less self-possession. Others, though, found it confusing.

Next to '—' sits Logan, a professional poker player, whose family had insisted he return to ceremony after a six-month hiatus in which he had become unbearable. 'I didn't want to get too reliant on ceremony, so I waited till I had something new to show,' he explained at breakfast on day two, which was made by my friend Martin, a former chef at Little Chef. His high-calorie Full English with Jubilee pancakes commemorating our Queen didn't go down so well in the keto-vegan-orthorexic company. Still, it came up surprisingly easily that same evening.

'Treat the ceremony like a table,' Logan told me. 'Slam down the full deck, hold back nothing, and it's rainbows and gold . . .'

'There are two pathways to legal usage in the US: either medical or religious slash spiritual, and I feel more naturally aligned with the religious one,' Oberon explained to me the morning after the first ceremony. The religious route, at least as he conceived it, was more free, less dogmatic than the alternative. 'The regulations around medical usage will always be necessarily stringent, applying all the different mechanisms of institutionalism: requisite training, education, certification, ethics committees, peer review . . . guidelines on handling the medicine, on dosage, storage; ensuring its composition meets certain standards – on and on and on, standardising every aspect of the sacrament's provenance, all in the name of research science or clinical treatment. It just feels out of keeping with the spirit or character of the experience; its strangeness, its history, its beauty.'

Oberon is scientific by disposition, reasonable and methodical to a fault, neuro-atypically so, I surmised: he completed his first maths degree by the time he was eighteen. He's also well informed, keeping more than half an eye on the emerging experimental literature on psychedelics. But from his perspective, subduing the 'strangeness, history, beauty' of the experience shapes the kind of research that gets done, and of all the psychedelics, he suggests, this is especially true for ayahuasca.

Although the evidence base is thinner for ayahuasca than for other mainstream psychedelics, there are encouraging preliminary findings for the treatment of psychiatric conditions like depression, eating disorders and PTSD. The work comes mainly from academic institutes in Brazil and includes trials with religions organised around the consumption of the medicine, such as Santo Daime. But even in such instances, Oberon believes, most of the studies neglect the huge fundament of indigenous folk knowledge and history surrounding the plants. 'In the last ten years so much of the conversation has been around Western models of therapeutic use, at least in the literature, because the people doing the therapy are the ones writing it. It's not that there haven't been exciting "findings"; it's the danger that what is being found is a reflection of what those who are researching are looking for . . . The medicines are converted into stable drug effects – the same way bio-chemistry understands other drugs – rather than the protean, unclassifiable entities they are. And the so-called "illnesses" of mental health are thought of as individual pathologies . . . rather than reasonable reactions to the state of the world.'

Ayahuasca's group setting represents a different model of mental health. Here, participation and relationality are active therapeutic ingredients. There are relatively few group studies with psychedelics, but these often point to a renewed sense of connection or connectedness, to self, others and the world (or, to use the jargon, 'the psychosocial effect of perceived emotional synchrony in collective gatherings').

I had experienced my own gradual version of this over the previous two nights. Initially I had been determined to remain separate, my prejudices about the other congregants having formed with the instancy of a Polaroid – 'too rich', 'too thick', 'too beautiful', 'too mundane', 'too interesting', 'too-self-absorbed'. On the first night of ayahuasca those provisional diagnoses had been confirmed and deepened whenever their journey had interfered with or interrupted my own. Ayahuasca's heightening effects on the senses, combined with the group setting, meant the space connecting us became so profoundly 'shared' that it felt like its own organ, a skin in which everyone is housed. And so, in the depths of the first night, when Too Self-Absorbed purged himself, his endless dry retches became two crocodiles with Covid frotting on the same parched river-bank where I lay, I took it personally.

But then there was another developmental stage: the dawning realisation that the suffering of Too Self-Absorbed was really *my* suffering: we were agonised crocodiles together until we were the same crocodile. I shifted from judging to identifying to becoming my neighbour: 'I *am* Too Self-Absorbed.' Forgiveness, compassion towards everyone including myself, were the only ways to flood this arid river-bank of entanglement, the only way, but also the perfect way. (In the midst of one's fantasy about Too Self-Absorbed, it's easy to forget you're most likely the subject of someone else's, which may be far, far worse than dry-humping reptiles.)

The final stage came the next morning when, after a muted post-ceremonial meal and a sleep as short and gossamer-light as a ninety-year-old's, I saw Too Self-Absorbed in the clear light of day: naked, human, by the hot tub, both of us ebullient in our happiness. Knowing atrocities have been perpetrated against one another and then healed, that what we're feeling now is the relief of survivors who are also murderers, we find ourselves hugging tightly, no holding back hips, not even the hips of an uptight Englishman, eyes singing '*but these are my real feelings*'. All this is part of the so-called 'art' of drinking ayahuasca, of learning how to maintain an open, curious, non-reactive attentional stance, of building community (albeit an overly fawning one) while undergoing a variety of unimaginably intense experiences.

Returning to the question of mental health, Oberon sees a broader problem. 'The standard medical approach doesn't allow for anything like healthy people getting better, for what we might call flourishing.

We have to position people as unhealthy in the first place – which frankly I'm happy to do because my definition of health is significantly more developed than the common one. You aren't just healthy if you are free of disease, or if you are healing or perfecting the self in some psychotherapeutic way. Let's set the bar high and say: you are healthy if you wake up in the morning, perhaps most mornings, and you have outgrown the idea of being a separate self altogether, you see instead how everything really depends on everything else at the most fundamental level, so that you are actively wanting to give of your gifts in a way which makes you feel profoundly fulfilled, and is also received and confers something like balance and retribution in your community and the world beyond.'

This was clearly a much broader frame than the standard model of individualised psychotherapy. Oberon and Sonqo have outlined what mental and spiritual health might look like in their 'Ten Commitments', laminated (puke-proof) next to the altar this evening:

> *I treat all beings with respect.*
> *I honour our traditions and those who came before me.*
> *I am of service.*
> *I pursue a life of virtue by living with integrity.*
> *I celebrate the natural world.*
> *I open my heart to the guidance of the divine . . .*

The commitments aim at being non-dogmatic, ethical, spiritual rather than religious. It is left to the individual to include any notion of the divine or not. And for all the eccentricities of some of the congregation, there are also many seemingly solid, grounded folk with years of dedication to the practice, who are caring and thoughtful with those around them, who know all the songs by heart, and who seem comfortable and confident in the depths of the ayahuasca night.

Of course, history shows that calling something a 'church' is no guarantee of good practice. As I write this, lots of people are looking to found churches across the US as a way of seeking a pathway to legal protection. One lawyer in Texas reckons to have helped register fifty psychedelic churches in the last few years. Some, like Sonqo, are established in good faith; others may be a path to making money on the black market.

Behind Oberon is an array of musical instruments, including a giant suspended gong, two metres in diameter. 'Tonight let's make it our practice to attune, to soften, to become more fluid in our receptivity to the insights that are offered to us . . .'

He holds up one of the murky bottles, muttering a prayer in Spanish which ends in a series of punctuated exhalations and glottal stops, like a poem of breath. Then he pours a cup of a dark brown liquid from 'No. 16' (each batch has distinctive qualities, known only to Oberon by its number) into a small silver goblet and holds it aloft. '*Gracias, madre ayahuasca, gracias chacruna, tabaco* . . . *Gracias Maestra Christina, gracias Don Rios, gracias a todos* . . .' Oberon salutes the room: '*Salud y buena pinta!*' To your health and wondrous visions!

He downs the cup, then invites with characteristic precision those who 'believe that they metabolise slowly' to step up to the altar and drink first. First up are the girl and her dying dog.

Mary, Oberon's mum, was introduced to ayahuasca alongside his father Tom when they were both in their mid-sixties. That was ten years ago. Oberon had been invited to build a home on their land, having spent a decade away, learning his *trade*; they thought he was a carpenter. When he arrived back home, he made a calculated decision to come clean to Mary, a general nurse, and Tom, a more or less teetotal economist who ran a large public policy consultancy. Such was the strength of his argument about the transformative effects of the sacrament that over the last ten years Tom and Mary have sat in more than 200 ceremonies led by their son. They have also allowed him to put his carpentry skills (he really was a professionally trained carpenter as well) to use in constructing a forty-berth temple in a bowered corner of their property, at the foot of Oberon's own newly built home where he lives with his wife and their young children.

Tom has spent the last two years of his retired life consulting for Sonqo's board, helping them with his vast professional experience to navigate decision-making around revenue streams, the management of the farm in Hawaii, ministerial salaries, legal concerns, outreach programmes, equity and diversity agendas and the like. At present Sonqo's fundraising is internal, with those who have more resources supporting initiatives that include free or assisted places for those who can't afford ceremony.

Not so long ago Oberon was invited to the boardroom of a venture capitalist looking to invest in psychedelic therapy, with the offer of a large consultancy fee. He turned it down. 'I'm not a therapist. I'm not a clinician, or a neuroscientist. And I'm certainly not an investor. What I am is a facilitator, a student of the plants, the carrier of a lineage. I cannot guarantee any type of experience or change to personal well-being. I can serve as an intermediary of sorts, but it's for the individual and the group to have the conversation with the plants and discover what, if anything, is of benefit for them, according to their own criteria. That kind of free-wheeling is not so sexy for business or for medical science.'

That morning after the first night of ceremony, I asked Oberon to tell me about the story of his own first encounter with the medicine. He explained that for the entirety of his childhood he had had no thought of what he wanted to be. Growing up a bookish, musically inclined boy in the Pacific North-west, he turned his back on a career in tech or academia because neither matched his appetites. He travelled first to Europe, then to South America, making his living as a street performer – magician, fire dancer, busker – studying each meticulously in turn, combining uncommon hours of practice with natural-seeming talent, until each competency flourished to his satisfaction. He also showed an uncommon propensity to tune into the 'bigger picture' of what he was doing, a gift he would later associate with the medicine itself. For example, as a street performer, he thought a lot about the nuances of how his act affected the audience. 'You notice it's not so much the technical competence of a particular act or rendition that makes for success; rather it's a matter of the performer's immersion in what they're doing that makes them compelling. You begin to see that most people have this craving for someone to draw them into a clearer sense of purpose and belonging.' As a clinician, I've certainly felt the same unarticulated need in my patients. It's also a recognition of power, of the potential of therapeutic abuse in the wrong hands.

A natural linguist, Oberon perfected Spanish and made significant inroads with Quechua, enough to allow him to perform local songs while busking, alongside staples such as Latinised Michael Jackson numbers and that gringo staple, 'Hotel California'. A virtuoso on classical guitar, he trained himself on various indigenous wind and percussive instruments. He became particularly known for his 'mash-ups' – braiding one song's lyrics to the melody and instrumentation of another, made possible by a deep underlying similarity in the grammar of each, like the two terms of a simile running in parallel.

> No one I think is in my tree
> I skip the light fandango
> And think to myself what a wonderful world:
> Strawberry Fields forever.

The Beatles to Procol Harum to Louis Armstrong back to the Beatles – a chain of associations afforded, according to Oberon, by the medicine itself: the right attunement, what Merleau-Ponty would call 'body-set', revealing things that appear to be concrete and separate as in fact malleable, responsive and interconnected at a more fundamental level. But at this point in Oberon's journey there was still no medicine.

So it went, for two more years, bouncing from one South American town to another, content with living as life unfolded, still without an acceptable answer to the question of what he wanted to be, other than just '*being*' itself, batting away the concerns of his parents, immune to ideas of 'losing time', 'wasting opportunities' or 'failing to get a foothold' – a home, a wife, a retirement policy – because he saw the deeper truth that 'there are no such footholds'; that 'life was only a steepling cliff face if your imagination obeyed the laws of gravity too'. Rather he preferred, like one of Hermann Hesse's carefree heroes – Knulp, Siddhartha, Steppenwolf – who he held in high affection, to hedge on deepening his sense of the present, living from hand to mouth, from thought to action. Until one day those actions *led* him to his first ayahuasca retreat. 'What can I say about that moment? To those that have been there, no explanation is necessary, and to those that haven't, none is possible.' Unlike many psychedelic enthusiasts Oberon doesn't let ineffability stop him. 'Let's say I saw myself and abilities, my life and its situation, from a new, unconstrained vantage, provoking in me compelling imaginings about what was possible, about how one might bring newness into the world, the hope of new harmonics in my powers of expression and what I might create . . .' He really does talk like this. 'But that's just flowery language, like saying I had to die or be reborn, or that anything can happen at any time; placeholders for this feeling of new capacities, alongside the richness of everything that was beyond my power, and the importance of honesty and humility for and the creative expression of one's gifts, and so on, and so on – only in infinitely more detail . . . Put it another way: one might say I didn't so much choose to be an ayahuasca minister, as much as being called by my future to become such. That in itself is just a theory, of time bending over backwards to meet itself. One might equally understand it in any number of no less compelling ways.'

Yes, but not quite equally, I thought. The idea of being groomed by a hidden fate in all the necessary requirements – prestidigitation, showmanship, musical skill, indigenous instruments and languages, speechifying, wisdom, empathy, a mastery of performance – to the point that you can reassure or persuade anyone of anything and mean it; then, regressing further back, the love of nature growing up, of wandering, of craftsmanship and storytelling: it might not be reasonable or logical, but that one story – of fate – had the power to eclipse any possible others because it alone carried the coherence, the rightness, the intelligence and surprise of art. Somehow – maybe it was just the medicine – my usual cynicism had fallen away. It was only a story, or it was a hackneyed New Age belief, or it was a mysterious principle of quantum physics. I was happy whatever.

It must be an hour since I drank the bitter cup. The temple has cracked open, the blackness a universe without limits. Somewhere in the deep recess of space Oberon is whistling a melody over and over; faint, complex, aspirated, his breath a bassline to the tune. I feel the melody on my skin, the crook at the back of my neck, a shuddering through the guts. The world – my body, its thoughts and everything thought about – becoming vibration.

Music lies on the fault line between the medical and religious models of psychedelics. In the brain, the classic psychedelics work on that bit of the neuron labelled the '5-HT2AR receptor site' which binds the neurochemical serotonin. The signalling that takes place here modulates the brain's response to music, affecting one's associations with meaning, mental imagery and personal memory.[1] In psilocybin-assisted therapy at Johns Hopkins, music is seen as part of the 'container': giving reassurance, facilitating peak experiences and emotional release, and providing directions in the absence of a road map. Much research recognises music as a powerful player in a psychedelic setting. For more thoughtful researchers, music becomes quasi-medicinal in its own right, by steering 'implicit learning' in directions that are most therapeutic. One researcher at Johns Hopkins estimated that music was contributing between 10 and 20 per cent of the changes seen in psychedelic therapy. But Oberon goes much further, proposing that the maths is almost perfectly wrong: the carrying of the sacred songs accounts for as much as 80 to 90 per cent of the experience, the plant brew the rest. 'What if music itself is the fundamental therapeutic agent that hides in plain sight?' he posits.

Wavepaths is a company established recently by one of the psychedelic researchers at Imperial, Mendel Kaelen. Its 'product mission' is to deliver the world's first personalised playlists to facilitate 'experience as medicine': 'We provide a new category of therapeutic tools, integrating psychedelic science, machine learning, music theory, psychotherapies and experience design.' The business is predicated on the ability of technology to 'instrumentalise' music, to give it a function that can be objectively measured and controlled. But as the neuropsychiatrist Iain McGilchrist notes, 'Music relies entirely on relations, on betweenness . . . is indivisible, ever changing, constantly evolving.'[2] It's hard, perhaps impossible, to measure music in any useful way because it would rely on finding an objective way to measure the richness of experience itself: it would mean quantifying a quality. When I described the use of music in psychedelic-assisted therapy to Jaron Lanier, his face screwed up as though he'd bitten a lemon. 'The idea of music as some catalogue of triggers is just horrible. Music is more alive, more mysterious, more precious than that.' It is also philosophically problematic: for Lanier, music is the only known thing in the universe that is its own context. 'It's so unusual – unique in that it's pure form without content and somehow it still works. It's hard not to sound a little mystical around it . . . but music appears to support something very fundamental, bound up with the nature of consciousness itself.'

Oberon's whistle has turned into a song.

> *Yachay sapa Pyka la lai la la lia*
> *Mamankuna dee dee dee*
> *Wio pacha manam conka . . .*

Three lines of Quechua about a medicine man who uses plants curatively, sung over and over again. On the recording afterwards I'll see the song goes on for nearly an hour. Right now it feels like eternity and a passing thought. From moment to moment I can't tell where it's coming from: the stone walls, inside me, or a tuneful bat in the temple rafters. Under the sound of people singing along someone is crying bitterly. I think it's Camilla, triggering in me a concerto of grief and identification, blaming myself for inviting her, then rationalising it as necessary for her healing, then blaming her for disrupting my journey, then feeling her tears fall down my cheeks; all of a piece, chained together by the one song. Later I'll learn it wasn't Camilla, and see in my storytelling around her fictional grief a parable for the greater part of our attempts to relate to others: we hallucinate much of our relational lives into being.

Oberon's lineage is Peruvian, like the more well-known Shipibo tradition. As with the Shipibo, it's a convention that the song's lyrics will describe a summons of spiritual entities, the causes or origins of the journeyer's problem, how the journeyer might be cured, cleansed or reconstructed, and the driving away of malign forces. The songs themselves are a 'sonic' substance which remain in the journeyer's body long after the ceremony. Oberon had studied for a brief time under the Shipibo during his time in South America. This meant a few *dietas*, retreating into the woods for periods of solitude during which various 'teacher' plants were ingested. The rationale is that in this way the healer will convert himself into a 'plant-like' being, able to communicate with the spirits in the journeyer's visions. But I've never heard him describe himself in this way: Oberon is a matter-of-fact facilitator, a simple carrier of the songs. He's more likely to tell me about the physics of music than he is about indigenous belief systems: how the syncopations, melodies, harmonies set up expectations, which like stories are either confirmed or disappointed. McGilchrist describes how these patterns lead directly to physiological changes, mediated via the right hemisphere of the brain's connection with the hypothalamus and the rest of the body, in breathing, blood pressure, temperature, the feeling of one's hair standing on end, and, of course, in crying.

The song comes to an end. The room falls silent a moment, the space becomes blacker still. Silent except for the lone sobbing. A bottom is needed and can't be found; just falling. Oberon explains that in situations like these the *icaro* threads silence, like a winch out of a swamp, like holding the line of a kite. A kite at night. A kite in a hurricane. A kite flown off the world's edge.

Oberon is knelt over the person, singing to her sobs, the songs the instrument of his care.

'The *icaros* are like a sailboat which you can tack according to the wind, by feeling the energy of the individual or the group coming through, a harmonising to sense and intuition. We choose songs that direct the energy.' A different planet from the prescribed play-lists of Johns Hopkins. 'The more one studies the harmony of music, and then studies human nature – how people agree, and how they disagree, how there is attraction and repulsion – the more one sees that it is all music.'[3] That's Hazrat Inayat Khan, Sufi vina player and esoteric musical philosopher from the early twentieth century whom Oberon is fond of quoting.

The developmental psychologist and psychoanalyst Daniel Stern thought that the first interpersonal relationship of all, between mother and baby, is best understood as a musical one, in the sense that it relies on the 'attunement' of the mother to the baby's feel-ings and her emotional harmonising with them. For the infant, 'the entire flow of maternal social behaviours can be likened to a sym-phony, in which the musical elements are her changing facial expressions, vocalizations, movements, and touches.'[4] Sasha Shulgin saw the different ways in which a chemical acts upon the person as being 'like the harmonics from the fundamental to the inaudible, which, taken in concert, defines the drug'.[5] He points out that with sound and musical instruments, we have ways of detecting, meas-uring, visualising these harmonics, but no such means exist with drugs. 'The true use of music is to become musical in one's thoughts, words and actions,' writes Inayat Khan. 'One should be able to give the harmony for which the soul yearns and longs every moment.'[6]

It is now the depth of night. Oberon has left the *icaros* behind to explore other instruments. There is the sound of an Amerindian double-barrel recorder that can throw its husky notes, a fourth apart, to any point in space, a melody so plaintive that I feel sadness itself, imper-sonal, ubiquitous, as it snakes through me. Then the cosmic drone of a didgeridoo. Briefly we are inside the bell of a giant swung crystal bowl. Equipped with hearing that is simultaneously drunk and bionic, I begin to wonder if I've ever actually heard music before. I no longer under-stand it as *sound*; there is no outside, no in, no subject, no object, no instrument, just sound becoming colour becoming childhood becom-ing space becoming the beat of the heart; musical notes as seagulls falling through lungs, or in the heaviness of bones. The philosopher Vladimir Jankélévitch saw in music an enchantment, or *charm*, that exceeded all the technologies of instrumental reason, really a species of demonic possession. The man inhabited and possessed by this intruder, the man robbed of a self, is no longer the sum of his thoughts; rather he has become nothing more than a vibrating string, a sounding pipe. The process borders magic more than empirical science, as though, following Jaron Lanier and Don Paterson, unlike everything else in the known universe, musical notes possess their own inherent meaning.

Right on cue, at what feels like the exact astrological apex of the night, as though inspired by the demon of J. Arthur Rank, Oberon smashes the giant gong and forty minds fracture into a billion bits which meet, intermingle, become each other somewhere in the middle of the temple space, never to separate again.

When I open my eyes Oberon, who even without intoxicants looks more than a mite alien himself, has become a fusion of human and praying mantis, playing an eight-string guitar – the extra strings allowing for additional harmonic resonance – with long, green, segmented arms. The image is somehow reassuring. He's closer to the plant, I think, the intercessory between us and it. But the very next thought is the opposite of reassuring: that he could skewer us with one javelin arm, prune our heads with a dry click of his mandibles.

Oberon takes up a new motif on the guitar: Latin, incantatory, stable, simple, calming, and slowly elaborates it, growing it into new variants in the petri dish of his capacities, as though he's watching and feeling something along with us, as though we are helping make it. It's not just that Oberon *sounds* great on ayahuasca, it's that he *plays* great on it (Camilla, a gigging country guitarist herself, describes him as having solo-level chops, but with jazz touch and a penchant for the flavours of world music). The tea fecundates his performance, suggesting possible elaborations, untravelled melodic paths down which to travel. For Oberon at least, ayahuasca is a performance-enhancing drug.

It's an Amerindian convention that the music performed in ayahuasca sessions tends to increase in improvisatory complexity, which is tied to the performer's extraordinary state of conscious-ness. A recent imaging study measured electrical activity in both audience members and performers during a classical music per-formance. The performers, classically trained musicians, were instructed to perform each piece of music twice: once in a 'strict' mode that adhered to a memorised interpretation, and once in a 'let go' mode which was explicitly instructed to be improvisational and spontaneous. The researchers found increases in 'entropy', defined by increasing complexity and unpredictability of brain activity in EEG recordings, in both performers and audience members during the 'let go' condition relative to the 'strict' condition. Researchers at Imperial have found similar profiles of entropy on fMRI imaging studies of subjects under the effect of classic psychedelics.

The research fits with my night, music bringing about an uncon-strained, hyper-associative mode of cognition that features unusually nimble changes in mood and in the way meaning is attributed. Ober-on's creativity syncopates with my own: with each key change I watch two rainbow-coloured, diamond-encrusted dragon heads turn in an instant to photo-realistic prize versions of themselves mounted on a wall, the fury of their death alight in their eyes, until a jauntiness in the rhythm makes them laughing Lego models of themselves. The unfolding perfection, the rightness of each moment, is staggering, as though what is being created through the music and sacrament, performer and audience, is an unfathomable work of art whose meaning, like life itself, can never be caught up with. This is music that is literally enchanting. It can't be straightforwardly 'observed' or gripped, another instance of what Merleau-Ponty calls a 'breakdown case', collapsing the distance between subject and object, a binding together of nervous systems that have been dysregulated, becoming creative by default. Unlike the research trials at Johns Hopkins, Oberon has no playlist for ceremonies; rather he sails into the night, responding in music to the complex weather systems and different ocean currents as he encounters them. And if he ever has even the slightest thought that he's playing well, playing by talent alone, play-ing 'for the crowd', that is, then the performance begins to fall apart.

Music has often been diminished by Western science. The nineteenth-century American philosopher William James thought it had 'no zoological utility',[7] meaning useless to all animals alike, I presume. The psychologist Steven Pinker reportedly considers it 'auditory cheesecake',[8] a hanger-on to language. And Sam Harris, neuroscientist slash philosopher slash podcaster, so important to me in terms of shaping my thoughts on different aspects of our culture with the acuity of his reason (of which more in the next chapter), has admitted that in his efforts to keep up with everything else he fails to make time for music.

If paying full attention to our experiences is a disposition, a practice, then for Oberon it's the combination of the songs and the plants – the music they make together – that is the guide for such attention. 'The songs and the plants that inspired them are truly the heart and the power, because they not only allow us, but *help* us, to go beyond concepts, limits, frames, attachments, and into mystery, possibility, infinity, the unspeakable.' It's the combination of brain-and-body-altering plants with logic-destroying music that is simultaneously destabilising and reassuring, so that the person can emerge from the darkness and discomfort exultant and inspired, albeit also somewhat bewildered. Everything is taking place within a kind of music: not just a metaphor for consciousness, but the mode of consciousness.

'Music is the key to the experience,' he says. By allowing us to proceed non-verbally we can avoid what he calls 'the tremendous sinkhole of talk therapy', which, important though it is, may only reinforce one's sense of one's individual identity, one's ego, one's de facto prison, and is therefore the very opposite of mystical revelation.

Oberon has stopped performing. The music has slowly migrated north over the course of the evening, towards a Western mode. The last hour is taken up with the slowly cohering congregation attempting to perform their own songs, like a very drunk version of *America's Got Talent*, only lacking the censor of a Simon Cowell. (I never imagined I'd actually miss him.)

As 'loose' as Sonqo purports to be, it's still a religious institution, and a Western translation of an indigenous practice. There are tensions that come with this: its doctrine of pure tolerance, jarring with the need for a kind of reckoning, to commit oneself to something specific (the etymology of 'religion' comes from binding oneself to belief); the vaguely cultic feeling of relentless positivity; the ambiguous extent to which the church is just a legal strategy. And then there is the question posed by Oberon himself: could Sonqo be maintained without him or would it disintegrate in his absence? There's the charismatic shepherd; then there's the flock: I cannot help but wonder to what extent the predominantly white wealthy congregation share Oberon's thoughtfulness, or to what extent they are in some sense ticking off the latest, whitest experience. Who knows how much they follow his example, his flexibility, his sobriety, his reasonableness, even about the limits of reason, or how much they branch off into their own creed of hedonism or New Age who-knows-what?

The magic is fading, but I am still exceedingly high, my reasonable mind hopelessly enchanted, more than a few degrees closer to the New Age. Over in one corner a Lebanese life coach roars like a dying animal; Joshua – at saturation point on the dance floor – is a mash-up, half man, half tree, in a one-man mosh-pit. Camilla has been softly whispering 'God help me' for the last hour. The dog is running for one last time in his dream-filled sleep.

Feeling the need for silence, I walk across the temple as though on a slack wire above a dark forest glade of predators. The anteroom beyond the temple hall looks like an acute psychiatry ward, or an ER room, with one real ER doctor facing a wall, head in his hands, saying 'Why, Jimmy – why?' Around him figures in white stand still, facing in different directions, as though posing for a futuristic album cover.

Outside, the huge, bleached rocks of high desert seem to me a charnel ground for megafauna, as though whole generations of palaeontologists had gone on strike. Under a curved sky, acned with stars, there is no human noise to entangle with, nothing living at all, save for the raking white light of a surveillance helicopter scouring the stony border for the non-indigenous on American soil. They're not looking for me. A lone tree stands out, as if it's the last living thing on earth. For the first time in my life I am overcome with the urge to hug one. From inside I hear that Oberon, the maestro, has picked up his guitar once more, strumming what sound like the chords of Michael Jackson's 'Thriller', but with a faster, funkier rhythm, and his own parody lyrics set to the melody.

> It's close to miiiiiiid-night
> Dressed in white, gathering around
> In med-i-taaaaation,
> Sitting in a circle on the ground
> You start to breeee-eathe
> You're soaring through the cosmos on a serpent,
> You're feeling freeee,
> Neon jaguars dancing in your eyes,
> Rainbow de-liiiiight,
>
> Ayaaaaaaa-huasca nights!
> Your buckets over-flowing,
> And your face is on the ground
> Ayaaaaaaa-huasca nights . . . !

A mid-Californian night's dream, Oberon, King of the Fairies. I'm dancing the zombie dance alone in the desert to the cosmic sounds of the Church of Sonqo.

'The true use of music is to become musical in one's thoughts, words and actions . . . and if you look with an open insight into the nature of things, you will read even in the tree – the tree that bears fruit or flower – what music it expresses.' Inayat Khan.[9]

I have to hug that tree, to feel my nervous system bind with *hers*.

> And now re-leeeeease
> On condor wings your soaring through the sunrise,
> You start to siiiiiiing
> Cosmic light in infinite reflections
> Your re-sur-rection.
> Ayaaaaaa-huasca nights!

Staggering, arms outstretched, to touch another life, to celebrate this astonishing gift we both share, to feel the healing power of another sentient body before I re-take my seat with the others.

> Ayaaaaaa-huasca nights!
> Your heart just keeps expanding,
> in this . . . crys-tal . . . frac-tal . . . of sound.

Just as my fingers reach out to touch living wood, I notice a black silhouette of a small man drawn on a yellow metal plate nailed to the side of the trunk. There are symbols of thunderbolts and then, 'DANGER 50,000 VOLTS'. The last words I ever read, had I hugged it. Not a tree but an electric pylon.

The closing ceremony is preceded by a share-circle in the name of integration and group cohesion. This is very much a Western adaptation. By contrast, in some indigenous cultures, including North American peyote rites, the privacy of one's experience on the medicine remains sacrosanct. Then there are the Brits, who, from centuries of enculturation, understand to a fault that not *every* story needs to be told. But then there's the other side: as much the circle palls, it's moving, connecting. Many have come through hell to get here; what from the outside seemed like a wrestle with death or insanity can now be smoothed into a different story, a rhapsody of revelation and growth, as though 'integration' requires a happy ending whatever the tone and content of the story that preceded it. Most are ready to step out blinking into the morning light, hearts basking in the 'aya glow'. A few are more subdued, which only punctuates their courage. Like Camilla, who, in her mid-sixties, clean and sober for three and a half decades, has travelled 5000 miles to give her soul seventy-two hours of dark night, just because she isn't satisfied with her peace of mind, is hungry for the possibility of deeper ease. When it comes to my turn I want to tell everyone how I have felt the plants playing on my cortex as though it was an infinitely vast church organ, making harmonies beyond my wildest imagination, showing me how my normal repertoire has been built on hitting the same few notes either side of Middle C over and over again with the touch of a toddler. I would have told them how, even long after the effects had worn off, I've continued to hear new music, strange and moving, inside myself. Instead, out comes clichés of my own, no different from the rest, back to those familiar, unimaginative, over-worn melodies. *I'm joining a church all over again.*

Oberon thanks everyone prolixly, gives each person a handout with Sonqo's Ten Commitments, flags up contact numbers for ongoing integration, points to a few CDs of non-ceremonial music of his own 'at literally no cost'. He has two lives: in this one he is the centrepiece of a community, a leader, a decision-maker, a minister, a performer, musician, minister of a church. But for most of the time he is a homesteader, a carpenter, a musical hobbyist, a family man known by his neighbours for having regular shindigs and camp-outs at weekends, and once in a while heading south to Hawaii, Costa Rica and – God knows why – Peru. Oberon knows there is a solution to the problem of two lives: sue the DEA successfully, legalise the ceremonial use of ayahuasca. But that would catapult him and possibly his family out of anonymity into media heat and light, with the potential for adulation or antipathy on a national scale. Not to mention the real possibility of mass demand: Sonqo's commitment to the quality and integrity of its offerings necessarily limits its inclusivity. Thus, that hypothetical solution to the problem of two lives involves the creation of another problem of two lives: although on certain occasions we may experience the wholeness and oneness of ourselves and the cosmos, our actual lives remain hopelessly split.

Outside the retreat centre I hugged everybody I could find, our atoms still mingled but less so. The helicopter continued its low flight over the desert, red and blue lights twinkling in the pale sky. I looked towards Mexico. It was time to excommunicate myself once more, from Sonqo, but also from the Western models of religion and science. It was time to head south.

6

Psychedelics and Meditation

'Just sit comfortably . . . and close your eyes . . . and let your body resolve itself into a cloud of sensations . . . Pay close enough attention so as to relinquish the form of your body, the shape of your hands . . . and back . . . and head . . . Just let each new sensation appear in consciousness . . . The mind is just a space where sensations are appearing . . . The moment you notice you are lost in thought, watch the thought itself unwind, and just come back to noticing the sensations . . .'

I'm listening to a guided meditation session on Sam Harris's Waking Up app. Meditation is, by definition, curious about experience, turning the spotlight of attention onto the workings of the mind, examining consciousness, the incessant parade of sensations, images, thoughts, from within. If conventional neuroscience gives us our minds in the third-person perspective, meditation gives us the same in the first person, bringing to our awareness the moment-by-moment perception of experience itself, of what it's like to have 'the lights on'. Metaphorically speaking, meditation invites us to widen our field of vision so that the frame through which we see everything comes into view.

In that sense, meditation is always doomed to fail. However *wide* we go, there will always be a frame. There can be no such thing as an unmediated experience of consciousness, because consciousness is the medium of all experience. The meditator's dilemma is like that of Archimedes: 'Give me a lever long enough and a fulcrum on which to place it, and I shall move the Earth.' Archimedes has a serious problem, namely no place to stand: he would need a second, larger Earth from which to perform his heavy lifting. The meditator and

Archimedes are de facto over-reachers, practitioners whose project hinges on a sleight of hand, like picking yourself up by your bootstraps.[1]

I am a Sam Harris fan. I often feel *relief* when I hear his view – on Covid vaccines, the development of AI, the insanity of political tribalism, the ethical opportunities afforded by charitable giving and other topics. Relief because contemporary reality can be befuddling, unreliable, lacking obvious ground. Sam is a 'reality instructor', to borrow Saul Bellow's phrase; to hear his reasoning is to watch a current norm or practice staked out under bright sunlight so the stains of bad faith might be seen. He often pulls at the premises of his target a thread at the time, until it has unravelled to a tattered heap of biases, distortions and bad incentives. Then, out of this unstitched material, he weaves something new with rigour and articulacy, so that what was confused or half-formed in me is made legible and sane. It's reason's own magic – somehow I feel both instructed and confirmed in my half-baked instincts, as he tips out the stone in the shoe of my mind. Relief, then, because at least one person has, amid the minefield, trapdoors and quicksand of contemporary culture, found a sensible place to stand.

A few years ago, Sam appeared to take a different turn, when he produced Waking Up. After all, meditation has conventionally been entangled with his bête noire, religion. But Sam brought his trademark grounded, searching, secular disposition to bear, framing meditation as a profoundly curious enquiry into the nature of mind and, ultimately, as a technology to enable human flourishing. The app draws deeply on a variety of Buddhist and Indian wisdom practices, but it converts them into Western-friendly modes of investigation. It promotes a common-sense, more-or-less effortless approach to mindfulness as a means of appreciating the richness and complexity of reality, the interplay of consciousness and its contents, and their implications for living well.

'You are the only person who knows how brightly it burns in your corner of the universe . . .'

Sam perfumes the end of the daily practice with a burst of poetry, because, for all its efficacy, reason only goes so far. We have this gift, he seems to be telling us, so familiar, so conventional-seeming, we scarcely recognise it. It even seems built to slip through reason's net. What the gift is exactly, that's hard to say. But poetry, meditation, music, psychedelics are ways of pointing it out.

'Thank you for your practice . . .'

No, thank *you*, Sam.

'—and I'll see you tomorrow for the next session in the Waking Up course.'

Yes, see you tomorrow. Best to Annaka and the kids.

Yes, for all his tics, I'm a Sam fan and thus, for my own sake, I need to resist him too. As anti-ideological and creative as he is, Sam is also part of a Western intellectual tradition that makes reason itself the only compass, its own ideology. This works brilliantly at disrobing folly, hypocrisy, bias, bad incentives. It may work less well when it comes to appreciating the value of the enigmatic, the creatively irrational, the unscientifically strange, affective and beautiful. Psychedelics, like music, poetry and spirituality, fall, at least in part, beyond rationalism. And just as reason is the only thing that can save us from many of the threats we face, it can also obstruct our appreciation of some of the things that make them worth saving.

This morning I'm doing my meditation practice in bed. After years of half-lotus, spine pulled straight, eyes open softly – embodying the dignity of serious enquiry, according to my teachers – this feels like a dereliction. The last two days have been solid travel, from the US to Mexico City, from Mexico City to Oaxaca, overland to Huautla de Jiménez for a Mazatec ceremony which didn't happen, so back to Oaxaca empty-handed, then a flight last night to here, Guadalajara, for a date with a new molecule: the Toad. I feel as though I've eaten too much sky. I don't know if it's all the tripping, but on top of a jawline-shredding chewing-gum habit I'm developing a thing for Mexican fizzy sweets. *Where the greater malady is fixed, the lesser is scarce felt*, is also the logic of secondary addictions.

I get up and look out of the window: grey-white apartment

blocks stretching without break to downtown, whose profile is lost in a polluted smear. No parks to break up the poured concrete, no chance of a green thought in a green shade, only a middle-aged woman in an Iron Maiden T-shirt trimming a small potted plant on the terrace opposite, four feet from mine. All the accommodation here is gated – no matter how little there is to gate – with attack dogs in the yard, many of whom look old, lenient, semi-retired. Last night, feeling the heat, I opened a barred window to let some air in for my evening meditation. Within two minutes my Airbnb host, whom I have never met and presumed to have taken off for the weekend, sent a text pic of the window with a red ring scrawled around it: as I was observing my thoughts, she was observing me.

I had been a fitfully committed meditator for more than two decades. It began in earnest with the two years I spent living with an Italian order of hermits. We practised something called 'centring prayer', a form of focused attention developed in recent decades by Father Thomas Keating from roots in an ancient mystic tradition founded on the thirteenth-century text *The Cloud of Unknowing*.

> When you first begin, you find only darkness, and as it were a cloud of unknowing. You don't know what this means except that in your will you feel a simple steadfast intention reaching out towards God . . . Reconcile yourself to wait in this darkness as long as is necessary, but still go on longing after him whom you love.

I was already an expert in longing; it was the waiting that was challenging. In addition to centring prayer, each morning between vigils and lauds the monks practised *lectio divina*, a slow, meditative reading of the day's scripture, savouring the words, allowing their suggestiveness to take on different sacred flavours. There was also the Jesus prayer, a form of Christian mantra from the Hesychast tradition of the Eastern Orthodox Church. And for the more liberal monks there was a thirty-minute zazen, a meditation from the Zen tradition, in the chapel rotunda after vespers. After years of chaos, in which I would snarl my atheism like a punk anthem ('the

Andy-Christ' was the name some friends gave me), I found unexpected peace sitting in old churches eating my words.

The monastery provided me with the beginnings of a meditation discipline, though less schooled in technique than I would have been as a monk in other spiritual traditions. After I left, my practice drifted. Later I picked up mindful meditation following Western adaptations of Buddhist teaching such as those of Joseph Goldstein, Jack Kornfield and Tara Brach. By this time I was working in clinical settings where 'mindfulness' was fast becoming the buzzword, associated with its potential for treating depression, anxiety and chronic pain. Those conditions were prevalent in the neurology department where I worked, but most of the emerging mindfulness protocols were thin on technique and assumed normal physical and mental capacities, making them difficult to adapt to our patients.

I had other concerns, too. This version of mindfulness was regularly portrayed as something like a technology – for reducing stress, for ironing out unhelpful cognitive biases. Instrumentalising it in this way often meant overlooking aspects of the Buddhist or Hindu philosophy that underpinned meditation practice, such as their concern with the actual nature of suffering, the transience of any experience, and the existence (or not) of a 'self', all of which had much broader implications for living than 'stress reduction' (and on which any meaningful, lasting understanding of stress might actually rely).

With my manager's approval, I took a year-long sabbatical in Asia to explore these traditions in greater depth, and research how various meditative practices might be used with different patient groups. Over the course of twelve months I made several lengthy retreats of up to two months long, beginning with different schools of Burmese Vipassana. In India I studied at the school of Yogananda in the northern state of Jharkhand, and then in the south under different teachers of Advaita Vedanta. In Nepal I attended teachings in Dzogchen and Mahāmudrā in the Tibetan tradition. In some of the lengthier retreats I sat for up to fourteen hours each day. Food was restricted, with nothing to eat after 10.30 a.m. Physical exercise, conversation, even looking directly at other retreatants, was forbidden.

Every action – walking, going to the toilet, taking the sleep from one's eye or scratching one's nose – was to be carried out slowly, deliberately, *mindfully*, as an extension of formal sitting. There was a ten-minute 'practice' interview each morning in which the experiences of the last twenty-four hours were scrutinised by an expert, and instructions for the next twenty-four hours were given. There were no distractions, or rather internal distractions became the objects of investigation. The whole environment was designed to expose the mind to itself in all of its chaos (even the retreat centre's walls were high). There was simply no escaping it.

Mindfulness in Western settings tends to focus on its becalming effects. In these Asian schools, though, it was often characterised by disruptiveness rather than tranquillity, by derangement rather than mental well-being. For my own part, I experienced some bizarre perceptual distortions. On one occasion I was unable to see my left hand for several hours even when holding it up before open eyes; rain drops on the outside wall were felt directly on my skin. Each day was an exercise in disintegration, dissolution and rupture, in overturning everything I 'assumed' about experience. Whatever the nature of a particular perception or sensation, whether ecstatic or, more often, unsettling, we were instructed to notice it and then let go.

The net effect of all this was a growing awareness of the habits and, through them, something about the mind itself: its tendency to find every aspect of experience unsatisfactory, to grasp at sensations and thoughts as if they were solid and durable, when the opposite was the case. Most unnerving was how this ate away at the concept of self. What was taken for granted as the stable, reliable vantage point from which I could consider everything else was, at the level of experience, neither stable nor reliable at all, but a mirage. Not only that: it was a mirage contemplating a mirage. Self and world were simply two points, two dots that met in the middle of a bridge over the precipice of nothing and, looking down, realised that there was no bridge, that the dots themselves were nothing too, other than a sense of falling, hurtling towards an imagined bottom, until there's the ambivalent kind of relief in discovering there's no bottom, in fact there's no falling, because there's nothing there to obey the laws of gravity in the first place.

The experiences that birthed these realisations can be understood scientifically as more or less invariant responses to a very particular, intensive set and setting – a mental laboratory, you might say. All the more reason to trust their validity, perhaps. What the retreats provided that neuroscience had not was first-hand, experiential knowledge of what I'd long known intellectually from the academic literature: that the 'self' is not a homunculus with a specific location in the brain, but the cumulative 'effect' of various different sub-systems. What we experience as thoughts and sensations are not the raw unmediated data of the world, but are made up primarily of self-generated padding, priors, predictions, conceptual filler. When that process of filling in is suspended or distorted, as it can be in meditation, then the data is suddenly experienced for what it is: fragments, prompts, next to nothing. And so we glimpse that what is taken for granted as a solid, stable object, including something as solid as our own body, is really a hologram, a trick generated by the compounding of perceptual slivers and innuendos. In such ways meditation makes the brain's patterns of integration and disintegration manifest: it is literally psychedelic.

In recent years neuroscience has taken a greater interest in meditation, in a manner which echoes its interest in and approach to psychedelics. This trend took a significant leap forward following the enterprising work of Daniel Goleman and Richard Davidson, who managed to convince a number of nonplussed Tibetan monks – some of whom had amassed the equivalent of several continuous years of meditation – to perform various cognitive tasks while wearing EEG-rigged swim-hats or lying in MRI machines. In the more experienced yogis they found that thousands of hours of meditating were associated with significant changes to parts of the brain's anatomy and connectivity.

Davidson and Goleman describe four principal neural pathways related to the range of meditation practices.[2] One pathway governs the brain's attention capacity. Of particular importance here were changes to an area (the reticular activating system) associated with the attention we give to habitual responses, especially physical sensations. In long-term meditators there was increased activity in this system, associated with the transformation of the habitual into the unfamiliar: the monks didn't experience 'shoulders', 'arms', 'hands', but the raw sensations that underscored them. The researchers also found increased activation in the area of the brain (the dorsolateral prefrontal cortex) which controls metacognition or 'thinking about thinking'. This corresponds with the yogis' practised ability to 'notice their noticing' and then let go.

A second pathway where changes were apparent (involving the so-called 'midline' structures) governs the sense of self, associated with the already mentioned default mode network (DMN), characterised by that ruminating, restless quality of mind that emerges when we are not task-focused. Like psychedelics, meditation appeared to change the connectivity in the DMN, reducing – to different degrees, according to concentration and experience – the mind's propensity to wander aimlessly, endlessly distracting itself, at the mercy of thoughts insatiably demanding their thinking. The third and fourth pathways where change was observed (based in the limbic system) are associated with the long-term meditator's capacity to alter his/her reaction to disturbing events, which accords with the generation of compassion and empathy. It's thought that these changes are associated with the long-term practice known as 'metta', in which the yogis systematically cultivate thoughts and feelings of well-being for every sentient being, beginning with the self and extending outwards.

From a neuroscientific perspective there is meaningful convergence in the fields of meditation and psychedelics. According to Chris Letheby, the Australian philosopher who collaborates with the research programme at Johns Hopkins, both lead to beneficial shifts in self-representation and what he calls 'cognitive dis-identification',[3] meaning we are better able to think about ourselves – indeed, to think about what it means to have a self – without blindly buying into the idle, often neurotic thoughts that the DMN typically generates. Both, therefore, have significant therapeutic potential in the field of mental health.

A few studies have explicitly looked at the two together: psilocybin and meditation are found to have similar effects on the connectivity of the DMN, which is unusually rich in the 5-HT2A receptors that are the preferred bonding site for classic psychedelics. Carhart-Harris and his team propose that the similarities are caused by similar patterns of increased 'entropy', the technical name for the kind of neuronal freedom also observed during musical improvisation, and which underpins the theory of the anarchic brain we encountered in Chapter 2.[4] Other research has noted that mindfulness meditation and psilocybin have similar positive effects on mood and social skills, mirroring hypothetically similar changes in neuroplasticity in the brain.[5]

There is also an emerging literature on how the two 'practices' might be beneficially combined. In one study, for example, meditators received either psilocybin or placebo in a randomised, double-blind fashion during a five-day mindfulness group retreat.[6] Psilocybin was found to increase 'meditation depth' and positive experiences of self-dissolution without concomitant anxiety. Compared with placebo, psilocybin enhanced post-intervention mindfulness and produced greater positive changes in psychosocial functioning at a four-month follow-up. The provisional signs are that meditation may enhance the positive effects of psychedelic drugs while counteracting some of the unpleasant effects associated with them.

Returning to an earlier theme, the apparent convergence of meditation and psychedelics may be reinforced by the way neuroscience sees them both. Stepping back, one wonders about the possibility that the overlap itself might in part be an 'artefact' – a consequence, in other words – of the way in which the science is deployed: that such similarities are progressively reinforced by the kind of attention, the kind of concepts or priors, the kind of vocabularies, the researchers bring to bear on their objects of study. Seek and ye shall find. In this way, two distinct though neighbouring territories – like a hand and an arm – become increasingly understood in each other's terms. And it probably goes without saying that in both fields, neuroscience pays far more attention to the activity of networks like the DMN than it does to the experiences themselves.

Sam Harris is also interested in the convergence of meditation and psychedelics. Waking Up is more than a meditation app. Through it he also offers other perspectives on and resources for the 'good life', including psychedelics. With their inclusion, Sam gives psychedelics a certification of sorts, proof that the drugs are 'part of the conversation', a conversation that is scientifically and politically literate, while being brilliantly and unremittingly *reasonable*. It's become something of a feature of his interviews that at some point, whether he's talking with a middle-aged nun, a bookish philosopher or a nerdy neuroscientist, Sam finds the opportunity, to ask, in a voice that's ever so slightly sheepish, despite its insistent maturity, whether his interviewee has ever taken psychedelics.

This morning, while the woman opposite continues to carefully prune the one plant left in miles of urban living space, I listen again to Sam's short audio-essay entitled 'The Paradox of Psychedelics', in preparation for the horror that's coming my way in an hour or two. Sam is addressing the overlap of drugs and meditation. 'Psychedelics prove to people that the mind is more vast, interesting, malleable than they expected before. A sufficient dose makes it clear you've been living in a kind of prison . . . '[7]

With the surveillance operation, the close quarters, the ageing guard dogs, this Airbnb also feels like a kind of prison. But in Sam's prison, the bars, exposed by psychedelics and meditation, are of one's own making, caused by the seemingly involuntary operations of the mind, and that hub of suffering – its guarantor, its jailer – the self. The unique strength of psychedelics, Sam suggests, is that, unlike meditation, its effects are both rapid and certain. 'Only wait an hour for a freight train of absolute significance to come roaring into the station of your mind and, for better or worse, you will get on it and, for better or worse, you'll never be the same.'

Certain assumptions underlie this claim. For one, it implies a high dosage. For another, it implies a certain degree of sensitivity (or neurotypicality) in the passenger: for some reason my monk friend Bede, despite taking a heroic dose, missed his train or was waiting at the wrong station. It also tacitly implies the absence of either a guide, a group of co-passengers, or a ceremonial setting. There is the further assumption of things never being the same either side the psychedelic experience. In recent weeks I'd come across several long-term psychonauts (indeed, even less seasoned pros like Frank) for whom sameness, in the form of repeated use – even allowing for the contradiction of it being the repetition of unusual experiences – appeared to be a defining feature; a species of psychological if not physiological dependence. ('Saturation point' was the Bitcoin retiree Joshua's term for it.) Medical science has always been keen to reject the possibility of psychedelic addiction on narrowly physiological grounds, but also because of the political and legal implications.

Caveats aside, for sceptics who doubt there's anything 'useful or profound to be had by observing the mind directly' (from meditation, in other words), Sam's psychedelic freight train may afford an 'easy access ramp' to a different perspective on consciousness and its contents, a first-person portal, if they're lucky, to a direct encounter with the realities that neuroscience has described from the third person. But for Sam, unlike the vantage point of meditation, psychedelics give 'a distorted sense of what it is worth finding there'. There is danger (Sam's tone gets more paternalistic, minatory even): the danger of being diverted from a possible encounter with a deeper underlying reality towards something that is random, self-referential and effectively deranging. The problem with psychedelics, as he has it, is that 'your emotional engagement with any arbitrary thing can achieve an intensity that has no reference point in ordinary life and is incompatible with it.' (Even now, a few weeks on and quite sober, I'm finding the litter bin at the foot of Yosemite Falls continues to cast its spell, a lesson in the depths of ordinariness.)

Harris goes on to warn that if you're coming from psychedelics to meditation, then you might think it affords you something similar: a way to reconfigure ordinariness. That would be a mistake. 'The true purpose of meditation is to recognise the freedom inherent to consciousness itself, whatever its contents.' In other words, while the science claims that psychedelics may free us from the tyranny of 'canalisation' – overlearned unhelpful behaviours – Sam suggests the drugs may contaminate meditation with a canal of their own: the insistence that everything must always be extraordinary. In fact, if the true purpose of meditation is, according to Sam, to understand 'what consciousness is like prior to our identification with thought, then you don't need a twenty-megaton change in contents to do that . . . in fact it might not be helpful.' Hence the paradox of the essay's title: psychedelics may be indispensable for some as entrées, but as to 'waking from the dream of separateness, they are *misleading and unnecessary*'.

Sam's rationale had some force, but in the light of my first five trips it seemed to me to lack nuance. It felt on the rigid side of 'reasonable'. I slumped back down on my prison bed, thinking that there was no known psychedelic that reconfigures ordinariness like the Toad, and you don't have to wait an hour for this hurtling train to arrive. Most reports compare the effects of 5-MeO-DMT to a faster, more visceral mode of transport, like being strapped to a rocket or fired from a cannon, and the experience begins in less than a few seconds, the moment the vapour touches the lungs.

Toad is the vernacular for naturally occurring 5-MeO-DMT. As previously mentioned, this extremely powerful, fast-acting, short-duration psychedelic gets its name from the most common source, the crystallised venom of the Sonoran Desert toad of northern Mexico. This is the substance that, according to Michael Pollan, blasted his 'I' to a confetti cloud, 'like one of those flimsy wooden houses erected on Bikini Atoll to be blown up in the nuclear tests'.[8] Though its provenance was uncertain in 2018 when Pollan wrote about it, at that point most people involved with psychedelics assumed that use of the Toad, like ayahuasca, had an ancient lineage. Ceremonies for its use were 'resurrected' by practitioners from the threads of circumstantial evidence, such as the toad-like figures sometimes found on Incan drawings. Meanwhile, celebrities championed it openly. Mike Tyson famously swore by it, in an interview on *The Joe Rogan Experience* speaking of the fifty Toad ceremonies that had turned his life around, and the public push-back in response to his profligacy.[9] Rogan was at pains to reassure him: there was no evidence of its harmfulness; he should do it as often as he liked if it made him happy. Well, Mike more than agreed: he literally bought the T-shirt, manufacturing a line of leisurewear with the banner 'The Truth Shall be Toad'.

Five years on there's been a shift in the Toad story. Hamilton Morris, celebrity chemist and psychonaut, dedicated an episode of his series on Vice to the likelihood that Toad consumption dates back no more than a few decades, making it more in line with synthetic creations like ketamine and MDMA, for which there is definitely no ceremonial history. Those that exist are modern inventions, more or less imaginative stagings of native rites. Meanwhile conservationists have raised significant concerns about the sparse numbers of Sonoran Desert toad, and the damage that the extraction method – milking the giant toads at the venom site in the neck – does to the animals. Medico-therapeutic applications have been compelled to rely on easy-to-manufacture synthetic analogues which avoid controversy and ensure the reliability of dosage. (Dosages of natural toad venom, where the 5-MeO is inconsistently distributed, can have significant margins of error.) This has created a factional divide between the naturalists and the synthesists, broadly falling under spiritual vs medical models, with both camps claiming different facets of moral and experiential high ground for their respective highs.

The last two years have seen an explosion of investment in researching potential medical applications of synthetic 5-MeO. For example, the large healthcare investor RA Capital has seeded two separate companies dedicated to developing 5-MeO-DMT for various mental ailments, recently ploughing $60 million into a Series A financing of Lusaris Therapeutics, which it incubated in-house. Josh Hardman told me that venture capitalists appear to love the idea of

5-MeO for practical and therefore cost-effective reasons. Unlike LSD, whose chemical trajectory through the body takes approximately ten hours, or psilocybin, preferred by the second wave of research into psychedelic therapies because it cut the length of the trip to four hours, synthetic 5-MeO delivers reliably self-transcendent experiences in fifteen to twenty minutes: roughly the same time it takes for a neurology outpatient appointment in the NHS. This changes the economics significantly, as Hardman explains. 'It's easier to slot into the existing healthcare system, significantly cheaper to facilitate and therefore an easier sell to payers.' For these reasons it seems likely that short-acting compounds like 5-MeO are likely to be a significant part of the medical future for psychedelics.

Bikini Atoll . . . Whatever Mike and Jo thought, Toad was no laughing matter; not to this terrified consumer. In recent weeks the ceremonies had been coming thick and fast. I needed to look after my mental and physical health, especially as I moved further from Western healthcare. I decided to engage the services of Aurora, my ayahuascara contact, for integration sessions after each journey via Zoom. We had become quite close friends in recent months, as I made my way slowly, haphazardly, towards a deeper appreciation of her viewpoint. She was pleased I was finally taking 'the medicine' seriously. And concerned I wasn't taking the risks seriously enough. To help mitigate them, I told her, my old friend Palmer had flown in from Kathmandu to be my trip sitter. Aurora thought that was a great idea. But then she didn't know Palmer.

Palmer, an esoteric 'seeker' by vocation, which he funded by programming in cyber security, had vast experience in the current field of enquiry. He was reliable and conscientious about certain matters, and even more reliably derelict about others. A dedicated psychonaut, he was unflappable in any and all psychedelic dimensions. But he had a psychological profile that bore the eccentric hallmarks of his dedication. If psychedelics do change your underlying philosophical beliefs, then Palmer's began changing before he hit adolescence; before, that is, more conventional beliefs could take hold. By the time he was sixteen he had a weekend job at Disneyland Florida, and when the park shut to the public he would dose himself with LSD and ride the empty rollercoasters on repeat into the early hours of the morning; his own personal psychedelic theme park. It meant that now, as he approached fifty, his relationship with reality was *different*, as different perhaps as that of the Tibetan yogis who spent their lives meditating in Himalayan caves (until they were transplanted to the metal caves of American neuroimaging labs). But Palmer's was a different *different*; a less conventionally healthy one.

There are many instances of this, but a recent one stands out in my mind. Palmer was, like the yogis, training in the Dzogchen lineage of Tibetan Buddhism, and devotees often base themselves in Boudhanath, the Buddhist World Heritage Site in Kathmandu, where they spend entire days repeatedly circumambulating the giant stupa reciting mantras counted by their fingers on prayer beads. Ur-psychonaut Terence McKenna, Palmer's hero, had spent a month doing something similarly tantric before heading out on his legendary 'Experiment at La Chorrera'. The Buddhists believe that a certain number – tens of thousands – of recitations and circumambulations are required to prepare the mind for receiving the pith instructions (meaning the 'quintessence of pure knowledge') of the guru in the advanced teachings of Dzogchen, holding the promise of instant enlightenment or *rigpa* – pristine awareness, literally 'knowledge of the ground of being'. The preparation phase usually takes years. Most never finish. But Palmer, being dispositionally lazy and an aficionado of cyber security, had developed a hack of his own. He rented an apartment in the ancient square overlooking the stupa. Inside it he reconstructed the stupa's giant white dome using a 3D printer, pasting esoteric wisdom texts on the model's facades, so that his living room was filled by a perfect miniature, ten feet in diameter and six feet high, leaving him just enough room to circumambulate it. Then, powered by mushrooms he'd grown by sowing fungal spores in the cowpats of urban cows, he walked around his creation reciting his mantras. In this way he calculated he could reduce the time it took for his purification by orders of magnitude, while saving himself from unsavoury encounters with the beggars, street hawkers and ill-tempered dogs who gathered round the real monument a few yards from his window.

The previous evening Palmer and I had been invited to observe a Toad ceremony at a temple in suburban Guadalajara, a recommendation from one of the Sonqo faithful. ('The shaman saved my life . . . again and again and again,' she told me. It wasn't clear whether she meant her current life on multiple occasions, or past lives in sequential order.) The shaman's chosen name was 'Krishna Sphinx' (while there's no I in team, there's a double shot in Krishna Sphinx) and true to form this Sphinx was a man of facets and enigmas. He was jockey-small, limber, youthful, at least from the distance of the low stools on which he'd seated us, below his modest throne. A debonair dark ponytail and Clark Gable moustache, oiled with essence of eucalyptus, were offset by loud clown trousers, rainbow-coloured necklaces and matching bracelets. His age, like his baptismal name, remained safely tucked away. (As I would discover, shamans generally don't like to tell you how old they are. It's part of the natural placebo: the implied elixir of youth slash the agelessness of their wisdom.)

Pencil moustache apart, his face was as clean and smooth as a child's, his eyes shone like a saint's with everything they'd seen, and the larger everything that couldn't be. But then the blood vessels around the caves of his nostrils had exploded, I assumed from *rappe*, an Amazonian powder mix of tobacco and palm ash that he self-administered with a small wooden applicator known as a *tepi* every twenty minutes or so. (Palmer was convinced it was really the snuff that made his eyes sparkle.) His language was as mosaic as his character: one moment a brain doctor, 'just like you' (tipping his cap in my direction as he proceeded to speak about complex anatomy and physiology, only *his* surgeries involved neither cutting nor 'scarification', depending only on pure knowledge of '*las plantas*'); the next, shifting effortlessly into esoteric ethnobotany, Amazonian folklore, Egyptian myth, New Age astrology, alien cultures, before circling back to 'Western ideas' such as 'quantum nutrition' and 'cellular gymnosophy' (both beyond even Google). It was as if to say, 'All traditions, ancient and new, all belief systems, medical and religious, all sartorial fashions, all ages, from childhood to senescence, have met in me and made their home.'

His necessarily humble wife sat on the other side of the altar translating, in between her own *rappe* habit, the shaman's words, a mix of French and Spanish (Sphinx was, we discovered in a rare moment of authentic biographical revelation, actually born on the Côte d'Azur), except for the occasional descent into English, when he wanted to make really sure we got the message. This polyglottery had the apparently desired effect of rendering her more cypher than translator: such was the mortal power in the shaman's words, it implied, that they could not be received directly. The wife – we never learned her name – though probably much younger than Sphinx, looked rinsed out by comparison, as though her husband was running his performance off her battery. I noticed how on a few pointed occasions her translation began before his speech had quite finished, or went on for a little longer than it felt it should: little revenges, I assumed, for a life spent in the shadows of the master.

As with other psychedelic sages I encountered, Sphinx's storied offerings had the feel of having been swallowed many times before, their genre unashamedly mythopoetic. 'That night marked the end of his childhood . . . He learned many things in the jungle, impossible to describe . . . The master told him he had never had a student like him before . . . Later it was time for him to go his own way . . . That night he died many times . . . The animals saw something in him that could not be mistaken, though it may be frequently overlooked by mere human eyes . . . He died again . . . He stayed like this for many days without moving . . . Death came easily and abundantly during that night . . . At one point – I can't talk about it . . . There are no words, except to say, I was once invited to serve Toad to the Queen of England – no, I really can't talk about it . . . *La reine incroyable . . .*'

Like her grandson Prince Harry, I thought, blue blood was temperamentally drawn to variants of DMT.

As was not the case with many of his claims, I thoroughly believed it when his wife told me that they had both taken one or other medicine in ceremony every day without fail for the last decade. Among the ads on Sphinx's website (for weight management, cellular nutrition, healthy bones, life hygiene, stabilisation and endonasal reflexology) were more or less credible-sounding testimonies as to the transformational effect of his work. Most memorable were those from patients who had received Sphinx's 'holy grail', pioneered by his teacher, an elderly shamanic 'genius', which consisted – to Palmer's and my astonishment – of a schedule of seven psychedelic animal and plant compounds (*bufo alvarius*, tepezcohuite, *salvia divinorum*, hashish, ayahuasca, peyote and yopo), delivered in a precise order and quantity at fifteen-minute intervals over the course of a single ceremony. Its effects met with international acclaim: '*nectare magnifique*', '*los siete magnificos*', 'ultravision', '*meraviglioso*', 'We have found Eden . . .' Anyone who undertook the regime would find themselves, Sphinx's wife told me, 'in full consciousness, full connection with everyone and everything'.

Full consciousness? I thought. A coma seemed more likely.

'Wowzers! How long does it last?' Palmer asked the wife.

'Three whole days. Minimum.'

'Is that safe?' I asked.

'It's halfway to heaven; three days, like Jesus in the tomb.'

Dead, you mean, I thought.

But Palmer was salivating. 'Is it still on the menu?'

As I looked censoriously at my friend, Sphinx and his wife looked at one another with touching regret. 'The creator was a genius,' she explained, 'but we started to feel toxic in our bodies. Our brains were requesting more and more . . . We went back to the creator and asked him. Eventually he admitted his sin: the holy grail had MDMA in it, and some amphetamine, a little LSD and cocaine . . .'

'It was not *organique*,' Sphinx chimed in. 'Not born of *las plantas*.'

'We stopped serving it immediately,' his wife added.

Towards the end of *How to Change Your Mind* Michael Pollan stresses the need to devise new ceremonies to fit our times. Imagination, ingenuity and sensitivity to the character of the medicine will be required to create culturally appropriate settings, he argues, and thus the right frames for meaning-making, alongside the need for greater thoughtfulness when it comes to the role of the guide or facilitator in Western usage. One can see how this would apply equally to medical and religious models of practice. It was definitely the case for Toad, where there was probably no historical practice to serve as a handrail. In the case of meditation, expert teachers are on hand to give regular guidance in technique, posture, diet. Many of these teachers have been schooled in Eastern practices for many years, and have learned how to adapt them for more psychologically minded Western yogis. In this regard our psychedelic ceremonies lag a long way behind.

Following the interview, Palmer and I had front-row seats for Sphinx's 'Toad Ceremony Magnifique'. His patient, a middle-aged Russian woman, had travelled halfway across the world to Guadalajara in search of 'healing after many traumas'. This, she explained, was her twenty-fifth consecutive night of receiving the medicine. During the course of the last four weeks she had shifted her sense of herself from 'patient' to 'student', the purpose of her ongoing 'work' to cultivate 'greater intimacy' with the animal poison, so that she could offer it to the 'many who need it' when she returned home. This rapid progression was not uncommon in New Age psychedelic circles: Oberon had told me that people often felt 'called' in this way, whatever the judiciousness of the calling, but the Church of Sonqo always turned them down.

Palmer and I sat to one side of the *maloca*, a small, low-ceilinged room four times as long as it was wide, with a mattress placed in the middle of the floor. The carpet was a deep violet plush, the walls spattered, Pollock-style, with brilliant fluorescent paints. It looked more like Syd Barrett's bedroom than a temple. The Russian woman took up her place on the mattress. Music began, a fusion of Sanskrit chanting and New Age electronica. Standing above her, Sphinx fired a blowtorch under the bowl of a glass pipe, vaporising crystallised venom until a narrow six-inch glass column filled with milky smoke. The woman sucked hard from the pipe, then passed it to Sphinx, who inhaled his fill and passed it on to his wife. Almost instantly the Russian woman's eyes whitened, small billiard balls that rolled back into her head, and she fell back limply on the mattress as though unplugged from the mains. A moment later she sat back upright with electric suddenness, held the position for a moment or two, then collapsed back on the bed. This same miniature sequence repeated itself over and over like a glitch on CCTV footage, the direction of play uncertain. Meanwhile Sphinx circumambulated the mattress perimeter, chanting, waving one arm high above his head as though conducting a meteorological symphony, the other holding a bucket into which he made petite, erotic-sounding purges. In between he swigged on a bottle of aguardente, which he spat out forcibly in the direction of the ceiling, spraying the room in clouds of 90 proof alcohol like a miniature Scotch whale. Sphinx's wife danced around her husband in circles, arms rippling like the goddess Kali, until she fell to her knees and buried her head in the belly of the Russian woman, a move she later described as an 'intuitive abdominal massage'. 'Jupiter and the dying moons', 'Jonah, nostalgic for his whale studio', were two titles Palmer and I considered for this avant-garde three-piece. There was more dancing, jaguar noises, feather-fanning, speaking in tongues. *Shaman-nigans*, is Oberon's term for it.

I cannot speak to the ceremony's intended meaning. Asking Sphinx to parse it afterwards, I got back gnomic fragments of Eastern lore and Amazonian cosmology, but mainly New Age platitudes and nostrums. Still, it was 'imaginative', at least, in keeping with Pollan's injunction. Maybe it was the fish-eye view from the back of the low-ceilinged room, the deep red of the carpet, the CCTV-glitch movement of the Russian woman, the bizarre choreography of the old, alcohol-drenched man with the Clark Gable moustache staggering around muttering obscurities with a puke bucket while his wife appeared to try to climb into her client's body in a surreal reversal of giving birth, but the whole thing reminded me of nothing so much as a sequence from a film by David Lynch, the drama at once strangely legible and utterly enigmatic. Was this really what Michael Pollan had in mind?

The ceremony's closing moments were a masterstroke of cinematic surrealism. The wife and the Russian woman remade the bed, and were meticulously removing from it with thumbs and forefingers numerous imaginary hairs, Sphinx pointing out 'the ones' they had missed. I imagine for them it was a training in scrupulousness, a co-ordinated act of purification and love, but what it looked like from the third-person perspective was a perfectly shared delusion, a *folie à trois* under a Toad's spell, or a dumb allegory of the power differential between a master and his acolytes.

Meanwhile the clinician in me was increasingly concerned about the shaman's neurological health. Working with the Toad at this frequency must have some physiological effects. Yes, he was a self-described doctor, but doctors are notoriously overworked, and unlike Sphinx most of them take weekends off. I thought too about the recent trend towards commercial investment in the medical application of 5-MeO, and the need to create safe and appropriate therapeutic environments. Sphinx's ceremony was certainly a bizarre outlier, but even so it begged the question: how could one shoehorn such a potent, idiosyncratic and ineffable experience into the confines of a clinical trial or a therapy manual, let alone a busy outpatient clinic?

5-MeO is known as the 'god molecule' because it can cannonball the user into a self-transcendent realm. This is white-knuckle 'ego-death' if, like the Imperial team, you're of a psychoanalytic persuasion (Robin Carhart-Harris holds a masters in psychoanalytic theory), or an insta-mystical experience if, like the Johns Hopkins team, you're of a more theological bent (their Principal Researcher, Roland Griffiths, is a long-term meditator with openly spiritual inclinations). At the heart of both concepts is the suggestion of the instantaneous abolition of selfhood, the psychedelic equivalent of enlightenment in a bottle. As Chris Letheby writes, 'whether there can be conscious states, induced by psychedelics or otherwise, that completely lack any sense of self', is 'one of the most philosophically controversial issues'.[10] Nonetheless, as he acknowledges, such experiences are a consistently recurring theme in reports from users of the short-acting psychedelic 5-MeO-DMT.

Philosophically, these are conventionally described as experiences of 'non-duality', meaning they involve a collapse of the distinction between the perceiving subject – the person's mind – and its object – whatever it is that that mind is apprehending. Letheby handles the concept of non-duality circumspectly. Like many spiritual seekers, the medical literature tends to talk about non-dual experiences as if they were uncontentious, even self-evident by some criteria. But the history of non-dual experiences, as mapped by various religious-spiritual, philosophical and literary traditions, is anything but straightforward.

A few weeks before arriving in Mexico I had read *One Blade of Grass*,[11] the memoir of Henry Shukman, a poet and Zen teacher, and one of the guides on Sam Harris's Waking Up app. As a disaffected younger man, as yet *naïve* with regard to meditation or psychedelics, Shukman travelled for several months around South America. One day, bored, lying on a beach looking out at the Pacific Ocean, he found his gaze drawn into unexpected depth and intensity. His account of what followed was so discombobulatingly profound that writing about it afterwards necessitated a switch from the first to third person, as though the experience itself was too bright, requiring a certain fictive distance to contain it.

The water was fascinating, blindingly white yet completely dark. Scales of brilliance slid over darkness, so it alternated between thick matt black and blinding light. But water was transparent, so was air, yet there the surface was, the sea's skin, thick as elephant hide. What was he actually seeing? As he pondered this question, suddenly the sight was no longer in front of him. It was inside him. Or he was inside it, as if he'd stepped into the scene and become part of it. He could no longer tell inside from outside. At the same instant the whole world, around, above, below – the sand, the sea, the light on the water – turned into a single field of sparks. A fire kindled in his chest, his fingers tingled, in fact everything tingled. The fire was not just in his chest but everywhere. Everything was made of drifting sparks. The whole universe turned to fire. He was made of one and the same fabric as the whole universe. It wasn't enough to say he belonged in it. It was him. He was it. The beginning and end of time were right here, so close his nose seemed to press against them. Suddenly he knew why he had been born: it was to find this. This reality. His life was resolved, the purpose of his birth fulfilled, and now he could die happy. He could die that very night and all would be well.[12]

Was this a 'non-dual' experience? For all the collapsing of subject and object there's still something separate doing the noticing. But that may be an artefact of having to describe what happened. Taken at face value, what Shukman describes is surely non-dual, and the suggestion of this passage is that somehow he chanced on a 'technique' that gave rise to it: with the right kind of quiescence and sustained mental fugue, call it relaxation or an entranced passivity, a more authentic relationship to the nature of reality discloses itself.

Such epiphanies are a commonplace in mystical literature and in Romantic poetry; Wordsworth's 'spots of time' are one example. Decades later, in the midst of formal Zen training, Shukman understands this earlier moment as an instance of *kensho*, Zen's term for self-dissolution. He explains *kensho* like this:

Imagine a pane of opaque glass. A hole is driven through it, and suddenly we see that there's a world on the other side of the glass: that's *kensho*. Koan study seeks to enlarge the hole, and create new holes, until over time the whole pane becomes riddled with holes, small and large, loses its structural integrity, and collapses. Then the separation between that world and this world is gone.[13]

Such experiences are often characterised, if not defined, by their inexpressibility (it's a criterion on Johns Hopkins' mystical experience questionnaire). After the previous night's ceremony the Russian woman had told us that 'to find right words is so hard' (though that didn't stop her from trying, at length). There are threads of neuroscientific theory that help understand ineffability – the quality of being beyond linguistic expression. The podcasting neuroscientist Andrew Huberman reminds us how 'downstream' or secondary language is compared to other cognitive processes,[14] which might make it hard for description to 'catch up' with overwhelming perceptions and sensations. Iain McGilchrist suggests that the non-verbal right hemisphere of the brain is responsible for processing complex emotionally bound data rather than the loquacious left hemisphere.[15] Of course, there are other ways of understanding reports of ineffability, too. A recent study of long-term meditators in the Vipassana tradition suggests they are a strategic way to preserve certain experiences in the memory.[16] In other traditions, they are a way of signposting God. In poetry, they may serve to underline the special credentials of the poet herself.

With Shukman's account of *kensho* it was the effability that I found so rich and moving, so moving in fact that I wrote to him; my first-ever fan mail.

Dear Mr Shukman

It's hard to put into words what your attempt to put into words something so ineffable meant to me . . .

I tried again a little less stiffly.

Dear Henry

Please allow me the chance to voice the effects your mysterious dumbfound-edness languaged in me on reading . . .

After years of going solo, with Waking Up my only handrail, Shukman's memoir had reawakened in me the wish for more intensive meditation training under an experienced instructor. Specifically, I was interested in koan training, long and arduous though it is, because it feeds off the conceptual limits of language.

What is the sound of one hand clapping?
What was your first face before you were born?
If you practise sitting as Buddha, you have to kill the Buddha.

Like clichés in reverse.

Such gnomic questions or paradoxes are taken into meditation, the idea being that in the right set and setting, with the right instructor, language can become the means to language's end and the arrival in non-dual realms. So at the end of my incoherent, grateful letter I smuggled in – as I imagined many other Waking Up devotees had – a gentle enquiry about his availability as a personal teacher. Henry was courteous enough to write back a polite, encouraging note, understandably protective of his time, but promising to send me a koan to sit with at some point. That had been several weeks ago, and I hadn't heard from him since.

I had encountered the lineage of non-dual experiences in the mystical traditions of different religions. My monastic teacher Father Bruno would often speak of 'unitive experience', or 'Christ consciousness', charting a tradition that reached back to the Christian Platonism of Plotinus and his notion of the One, or Heraclitus and his idea of phenomenal reality as 'flow' or 'flux'. The same thread could be found in the writing of later Christian mystics like the author of *The Cloud of Unknowing*, Meister Eckhart, St John of the Cross and twentieth-century writers like Teilhard de Chardin, Thomas Merton and Bede Griffiths. It was a commonplace in such writing to describe the sacredness of non-dual experiences 'apophatically', in terms of what they are not over what they are, an idea not dissimilar to Keats's 'negative capability'. Father Bruno would often describe the mystical tradition as being reluctant to create a rational framework by which to systematically engineer such experiences. While they required certain conditions or settings – monastic asceticism, for example – and a particular 'set' which was diligently sceptical about impostors and false gods – it really boiled down to a cultivated openness in personal disposition, a passivity that allowed the entirety of one's heart to be directed towards God. The mystery of experience would follow, if grace permitted it.

Later, when I came to Buddhism and Vipassana practice, I found a much more detailed system to engineer non-dual experiences. Vipassana, from the Pali for 'insight', is designed to root out 'the experiencer'. As with neuroscientific understandings, the 'self' is understood to be *'pariyatti'*, the Pali term for a thought-generated image or a conceptual overlay. At the intellectual level this is easy enough to understand. But the tricky part is seeing how much 'thought' is responsible for the world we live in. Even if non-dual experience is reaching the place beyond thought, everything we say about it is by definition the same old stuff of the thinking mind, returning us to the problem of Archimedes and his lever.

As I was only really a novice at Vipassana, clocking up two or three thousand hours at most, I consulted an expert. Michael Highburger and I had briefly been monks together, twenty-five years previously. Since then, Michael had relocated to India, spending the last two decades living in the ashram of Ramana Maharishi, the doyen of the neo-Advaitans. Michael looked after the Ramana literary estate and newsletter, which took him about six long days each month. For the rest of the time he spent an average of ten hours a day in meditation, rinsed and repeated over more than twenty years (he joked that he was trying to repent for his former life as an acidhead and philosophy student), and knew many of the Eastern traditions inside out. I wrote to Michael about my 'lever' problems in the non-dual realm, and he responded a few days later, the moment after stepping out of the latest shift in a lifelong retreat.

I have no advice other than, 'Be careful!' Theorising about non-duality invariably involves excess and, perhaps, a lack of ontological humility: in the impulse to make non-duality and the experiencer into something *graspable*, we make it into something *else* (and our narcissism may insert itself in the bargain). If, earlier, we had a small self that said 'I', now we have a small self that says, 'I know'. It is understandable that we might mistake this for true knowing, but the true knowing comes when the 'I' is removed altogether and stays that way, says the tradition. Because any 'I', even its ghost, cuts us off from direct experience.

Humility, or just plain innocence, was the defining characteristic of the teenage Shukman's moment of being on the beach in South America – the young Henry had no training in so-called 'spiritual materialism'. Rather what appeared to make his experience of non-duality possible was its trance-like thoughtlessness. To analyse it, to think about it, was to make oneself an agent all over again. And language was the chief culprit: all words signify man's refusal to accept reality on its own terms, to paraphrase Walter Kaufmann. The refusal is not just an intellectual defence; it's a psychological one too, because, as some researchers have speculated, our compulsive thinking – our over-developed DMNs – may, in an evolutionary sense, have been designed to reduce our anxiety about potential threats, persisting to this day when such threats no longer exist.

Viewed as such, discursive thought – even especially our reason – may be our main strategy for keeping what is existentially terrifying at bay, even when it's the very last thing that's called for. The Buddha tells the story of a warrior pierced by an arrow who, before allowing the field surgeon to extract it, insisted on knowing which wood it was made of, who its owner was, etc. According to Buddhism, as Highburger's email explains, 'We attach to thoughtfulness, to reasoning, to description . . . And in this way we make our wandering mental life real and solid despite it being illusory, which in turn becomes the source of our suffering.' So avoiding the temptation to make thin air solid, or thought a refuge, requires cultivating the psychological strength and spiritual courage to embrace the knowledge that there is no self, nothing substantial at all at our centre, that there is no centre.

One may have a range of profound, insightful experiences in meditation (including non-dual ones), but the issue becomes how to translate their wisdom into the rest of one's life. The natural tendency is to revert to one's previous way of being. According to the Buddhist tradition, whatever has been glimpsed in such moments cannot be incorporated and stabilised into new ways of behaving without years of practice or preparation. And even with such practice, without a commitment to ethical living (*sila* in the Vipassana tradition), or to self-purification (*nondro* in Tibetan Dzogchen), there is no foundation to root any 'gains'; real, sustained progress isn't possible.

What is true for meditation would appear equally pertinent for psychedelic experience. Yet the working model for psychedelic-assisted therapy, to the extent that one exists, is that a single session might be sufficient for enduring change. It seems to lack 'psychotherapeutic' humility, to adapt Michael Highburger's phrase, especially when it comes to non-dual experiences: how can you integrate an experience that cannot be conceptualised, that cannot be adequately rendered in language for therapeutic discussion?

Perhaps a starting point is to adopt the distinction between a philosophical self that somehow disappears in the non-dual experience – whether on psychedelics or in meditation – and the 'psychological self' that has to live with it, make use of it, develop or heal because of it, according to its limitations. Not to recognise this second self, as fragile, earthbound, 'clinging' to duality – either side of psychedelic 'therapy' – would be to fall into the hubris and grandiosity that soaked Sphinx's ceremony. He appeared to operate recklessly at times, he certainly lacked humility, and yet perhaps no man on earth had spent more time in Toad-inspired non-dual realms than the diminutive shaman from the Côte d'Azur.

Thinking back to Henry Shukman's early *kensho* experience, there's something about his relaxed passivity that seems to be a requirement of the 'set'. In the moments before, Shukman was lying down, listless, bored; there was no effort required in generating what was to come. In Sphinx's ceremony I had observed the Russian woman lie down in the moments after her inhalation. I wondered about passivity, and its relation to the psycho-spiritual wisdom of 'letting something go'. Phenomenologists like Merleau-Ponty believe that the object world has a life of its own, regardless of human action, regardless even of subjectivity. Viewed this way, non-dual experiences, in meditation, psychedelics or through poetic epiphanies, are really bringing us into a secret: the secret of how things really are (of which more later).

All of which might add up to saying that I had spent the entire morning jet-lagged, stupefied, lost in discursive, abstract thought, as a way of defending against the terrifying prospect of the imminent Toad ceremony with Sphinx. The meditation, staring at the woman trimming her plants, the speculations on philosophy and religion, were really little more than attempts either to distract myself from or rationalise what was coming my way. As a soon-to-be-human cannonball, I had no map, no seat belt, no airbag, and my 'chauffeur' (more accurately my 'explodeur') was straight out of David Lynch.

I had formed no 'intention' for the ceremony, but even if I had, what would be the point? You can't hit the brakes and the accelerator at the same time. There was a trickle of evidence indicating the benefits of meditation in psychedelic practice. But these were coarse studies, proofs of concept, perhaps, and had little to say about the kind of non-dual experience that might lie ahead. And despite huge recent investment, there was no research linking meditation, or any other nuances of set and setting, to 5-MeO. I might try a body scan, for example, but 'Where you're going,' Sphinx had told me, 'there is no body.' Most clinical applications of mindfulness were all really bound to dualism, predicated on a more or less stable subject (the self) training itself to be more or less aware of more or less stable objects. But what happens when the self disappears, turns to vapour like Toad crystals?

Without wishing to sound too West Coast, all I had was my 'self', and all that self had was its 'practice'. I resolved to meet the experience as a yogi, to bring Sam and Waking Up into the depths of psychedelic space: seated, dignified, humble, reasonable, a representative of science who is willing to surrender to realities that were beyond either my control or understanding, as if the Toad cared about any of that. Either way, these fragile ornaments would be my crutch, a hand held as whatever was left of me paddled at the shore of Bikini Atoll.

Just as Palmer and I were readying ourselves to leave the apartment, I got an email notification from Henry Shukman, replying after weeks of silence.

> Hey Andy
> If you do want to try out a koan, a great one is:
> 'What is this?'
> Which can be abbreviated to just:
> 'This?'
> HS

Just on time. Henry's koan would be the 'intention' I would take into ceremony.

'Sphinx says the medicine is a spirit,' his wife translates. 'It knows you very well. All that spirit gives corresponds to what you need . . .'

'This' is exactly what I need, I tell myself. Only 'this' . . . Meanwhile the rest of my psychological self is fucking terrified.

I'm sitting upright in half-lotus on the smooth, impeccable mattress, my breath slow and deep. Sphinx is holding his blowtorch to the crystal bowl, making his peroration over the top of his wife's translation and another track of New Age fusion music. 'The Toad *knows* all. The Master's actions and touches are precise, as precise as a brain surgeon . . .'

I take a large draw of thick white smoke and hold it deep in my chest. The taste is both sweet and septic. Instantly the space of the *maloca* fractures, the sense of time passing, of having a body, begins to chasm. Duality is crumbling like the masonry on a temple dedicated to Descartes at the foot of a volcano. I have vowed to hold my position; to look on tempests and not be shaken; to find the membrane of active and passive, between instant enlightenment and infinite struggle, between sitting upright and lying down, between self and object, mind and world . . . The trouble is that I have no actual position to hold, anywhere, nor an 'I' to do the holding. Goodbye.

What happened next was . . . unexpected.

It seemed that I had not quite broken through, that the dosage fell a little short.

Can I level with you?

I didn't trust him. I knew deep down that the setting was unsafe, so at the very last I held back, stopped myself from drinking the last sips of milky air.

It kept me divided, split, on this side of the non-dual realm. I retained the warped and dislocated sense of Sphinx staggering around spuming clouds of alcohol in the manner I had seen the previous evening, and of my own body seated on the mattress. 'I' was still here, in some fractured way.

At some point the shaman moved deliberately towards the temple altar, but as he went to sit down, he missed the throne altogether and slumped on the floor. With his head propped against the altar I watched through the maelstrom of tessellated space-time his shiny eyes roll back, his arms and torso begin to shake, his tongue loll out of the corner of his mouth. In other words, what was happening was not ineffable: I was present enough to recognise the physical signs of a partial seizure.

Sphinx's wife moved stroboscopically across the temple floor to her husband. She held him up with one arm, using the other to perform her speciality abdominal massage. Somewhere else the Russian woman, who had taken a dose as part of her ongoing training, was wailing so loudly the music could no longer be heard. Her cries liquefied into a series of violent purges, each of which missed the many buckets that were stationed throughout the temple.

This was *my* ceremony, I kept saying, though it was increasingly hard to feel central. I sat through it all: upright, dignified, a lone still point, unable to move.

Later Sphinx would explain that his seizure was for my benefit.

In fact, he would tell me, it was called forth by my brain not his; the Toad using him to *perform* my relationship with my father, or to manifest my 'bedevilment' (his word), symbolise my 'broken' connection to the 'life force', to bring forth deep childhood shame, or all of them at the same time, it wasn't entirely clear. Such was the riddle of the Sphinx. And as he spoke I thought again of Michael's stress on humility with respect to one's world view, and the lengths we will go to defend ourselves against the terror of our frailty.

While Sphinx was fitting I experienced a paralysis of sorts. I wanted to go to his assistance, but my body would not respond. Then Henry Shukman's koan came back to me: *What is this? This?*

It cracked open instantly. There was no *this*. 'This' – the word, the concept – pointed to 'presence' in all its strangeness, but even that concept was only a label. With a head full of Toad there was nothing to do the pointing with, no location to do the pointing from, and nothing to point at, *except* presence. Just 'this'; only 'this'. 'This' was both the non-dual experience *and* its articulation. I didn't need to do the twenty years of training in thousands of different koans. All I needed was an ethically managed Sonoran toad farm.

An hour later Sphinx was back on his feet, blowing *mimosa hostilis* (another plant-based DMT) deep into Palmer's brain through a blowpipe the length of a samurai sword. I left them to it and walked the two miles back to the Airbnb through the warren of gated houses and rheumatic guard dogs. For the next few hours, I remained suspended inside my koan, a prisoner of *this*. By the time I spoke to Aurora the next morning, there was nothing left of 'this' to report.

The *shamanigans* of Sphinx were clearly uncontained, borderline dangerous from a medical perspective. But then without any indigenous usage, and only a brief underground tradition to draw on, so little is known that could helpfully inform the taking of 5MeO. Besides, what could set and setting really mean in the face of instant non-duality? What could integration mean if there's total disintegration? The idea that meaningful change could be delivered by a single lightning strike, or hammer blow, and sustain its force over time, ran counter to most of the clinical therapeutic experience, where change was slow, hard-fought, incremental.

In Dzogchen, realisation is instantaneous, but stabilising it takes the active discipline of the rest of one's life. In Vipassana, insight is slow, accretive, multifaceted, with enlightenment a remote possibility. At the heart of both is the discipline of integrating the experiences of formal meditation into ordinary existence – which, alongside the maintenance of an ethical life, is where the real work is done. Meanwhile, the ancient models of plant medicine have an entirely different emphasis. There it is commonplace for the shaman to treat his patients by taking the medicine himself, and then directing the effects to the healing of the unintoxicated patient. The patient is active only in his imaginative reception of the shaman's journey. But in both Buddhism and the indigenous use of psychedelics, there is a whole world view shared by a community of practitioners, a long, evolving tradition in place to help make sense of experience. Compared to this, the medicalised concepts of set and setting, intention, assistance and integration are recent, earnest, somewhat naïve attempts to rebuild infrastructure on Bikini Atoll without any foundation underpinning it.

I returned to Sam Harris's paradox: that psychedelics may provide us with an escape hatch from the illusion of the self, a fast-track access to the non-dual, and yet because of the dangerous equivalence of all psychedelic experiences, it is entirely likely that other illusions will simply replace the ones we have shed.

I wondered how the paradox had been shaped by Sam's previous experiences – his 'set', that is. On different occasions Sam has outlined how several consecutive bad trips in early adulthood made him stop taking psychedelics altogether for many years, until a more recent heroic dose of mushrooms. From what he's said, such experiences were mainly informal, solitary and conducted outside any spiritual or medical model. By contrast, my set was something like the opposite: laden with a variety of contexts, literatures, guides and models. Though I was barely halfway through this increasingly challenging taster menu, the first six courses had already pointed to a whole array of potential ways to avoid that pitfall, and uncover new ones of my own.

Different compounds in different settings at different dosages had led to wildly different experiences. There were berserk freight trains like Sam's, for sure – look no further than my dinner with Sphinx – but there were other vehicles too, which moved at different speeds, allowed for certain choices to be made. Not all involved being cannon-balled instantaneously out of oneself. With some you might crack open a window, use binoculars to explore a detail of the scenery more closely, or sit back in your seat and chew on the authenticity of an insight, learn how to dance with your demons, stub your toe on the joke at the heart of a challenging experience. Even micro-dosing, which by medical convention was sub-perceptual, could become interesting – *perceptual*, in other words – if one brought a meditative attention to it. And all of this might be further shaped or elaborated by a guide, or with appropriate conversations with a well-disposed community of elders, versed in their respective wisdom traditions. And, I'm repeatedly told, if the experience is well cared for, it can retain its force in the months and years that follow.

With respect to Sam Harris, meditation can play a part in this. It may, if the dose isn't heroic, be useful on a trip, as a way of meeting the reality of such experiences with a degree of volition and equa-nimity. Also, with the right disposition, a bad trip can be re-framed otherwise, as being stuck in the birth canal of a deeper more rewarding experience, or whatever other metaphor might suit your way of expressing things. But that's the point: the experiences *are* expressible.

Six months later, as I write this, I see groups advertising on Whats-App for a meditation this evening at which the yogis will chew leaves of salvia ('No burners allowed'). Tomorrow night there's a holotropic breath session nearby with small doses of ketamine. People are finding ways of bringing the two realms, meditation and psychedelics, together in a variety of ways. Which is to say that Sam's paradox is personal, limited, only a sliver of the spectrum that might exist between the default 'mind prison' and a unified 'beatific vision'.

Come on, Sam. Don't give up on the drugs yet.

I might offer an alternative paradox of psychedelics. Although they may be beyond the grasp of rational knowledge, we refuse to stop flailing in their direction, trying to catch them in our nets of conception and reason, which themselves are woven out of specific notions of progress and improvement. And so when we *wake up*, or *make sense*, when we think we've caught them in our net, we also stand in danger of obscuring the inherent mystery of psychedelics with our generic (and yet overly personalised) models and hypotheses. We symbolically toast our success as if we have resolved a complex problem. Meanwhile, the real riddle continues to sit inside us undiscovered; sphinx-like you could say.

7

The Psychonaut

Psychedelic experiences are deft at giving models the slip. While science MRIs its medicine, and churches chug their sacraments, the actual psychedelic experience throws its homework on the fire and takes the car downtown to dig a pony – to mash up Bowie and Lennon – and we well know a pony can't fully be dug by reason alone. With their instrumentalising of the drugs as medico-mystical technologies, the different models stand in danger of straightening the kooky, of missing the most manifest characteristic of mind-manifesting substances, the rainbow-coloured elephant in the Johns Hopkins living room, which *is* their loopiness, their strangeness. Or to reverse the metaphor, if the models were themselves elephants, they would have overlooked the acidhead in the middle of the savannah; the lost-looking, uncamouflaged dude in the pink top hat searching for something tall to tell his tales to.

The neglect isn't some species of accident. The civilising raids on psychedelic experience were the obverse of the War on Drugs. In 1970, the new Controlled Substances Act classified LSD and other psychedelics as Schedule I, signifying that they had no medical value and the highest possible potential for abuse. 5-MeO might be the experiential equivalent of Bikini Atoll, but LSD was its cultural equivalent: 'The Most Dangerous Thing Since the Atom Bomb' ran the headline of Sweden's *Gothenburg Post* in 1967. The association of LSD with the overreach of counterculture is one reason why the current wave of medical research, with notable exceptions, has tended to steer clear of it. And for all the modish talk of psychedelic inter-disciplinarianism, there is one significant tribe who remain disenfranchised, who wouldn't want enfranchising if you paid them (unless you paid them in Pink Panthers or Orange Sunshine, or other notorious strains of acid), and whose personal vision of hell might be to end up in the 'control' group of a psychedelic randomised control trial; a tribe for whom there isn't any Psychedelic Renaissance because the drugs never went away; they just went underground. I mean the psychonauts.

It's strange to think that, while neuroscientists interested in meditation travelled halfway across the world to Himalayan caves in search of 'expert' brains to scan – the yogis who'd amassed thousands of hours of practice – their colleagues in the newly minted departments of psychedelic science were not interested in the men and women, likely to be living on their doorstep, whose brains had been exploring other dimensions for several decades. Imagine what these yogis' MRIs might look like: tremendous crystalline waterfalls of plasticky neurons, amygdalas in the form of machine elves, default mode networks run amok like self-dribbling basketballs . . . (Perhaps the most legendary of psychonauts, Terence McKenna, died of a brain tumour that apocryphally took the shape of his beloved mushrooms.) Meanwhile, a few miles away, medical research consults with itself on a different kind of interior design: the layout of their 'living rooms'.

There was an entire vernacular culture, a wisdom tradition of sorts, that carried on when the law ended the first phase of research in the early Seventies. Luminaries like Sasha Shulgin, the McKenna Brothers, Robert Anton Wilson and Dale Pendell, bringing their own wonky, undisciplined interdisciplinary polymathery, from fields such as chemistry, ethnobotany, physics, to bear on mapping out psychedelic space. In *High Weirdness*, the acidhead and social historian Erik Davis describes the different ways in which some of the aforementioned intellectuals deliberately put their own minds on the line in search of their secular versions of mystical experience.[1] Alongside grandiose summit-seeking questing, there was a compelling self-awareness: a scepticism about the meaning of any quest, an alertness to their own capacities for hyperbole and self-deception, and a pioneering courage to lose the path, to lose the mind even – sometimes for long stretches – before regaining it. This is the kind of loopy 'experiential-practitioner' model that's excluded from contemporary science and the research lust for 'psychedelic naïvety'.

Reading Pendell or Shulgin, or listening to Terence McKenna's nasal, incantatory sentences, makes clear that distinctions between psychosis and insight, between challenging, traumatic and mystical experiences, are not watertight, are prone to collapse, are at times hopelessly confused. In this realm, Davis tells us, spirituality includes profanity, as a check on the sacred. It's understood that the perverse, the playful, the punkish and, perhaps most pressingly, the humorous – with its built-in cruelty – may kick-start self-transcendence as much as listening to the hallowed chants on a Johns Hopkins playlist or some Western, Educated, Industrialised, Rich and Democratic (that other WEIRD) white guy singing in Quechua.

LSD, synthesised in the late Thirties, and coming into scientific and cultural prominence in the following two decades, is the perfect token and informant of this weird counter-history. Unlike mushrooms or peyote, it has no sacred history to draw on. Instead, acid alerts us to its profanity, its literal 'pulpiness', by the goofy comic images printed onto the blotting paper in which it is typically stored and consumed, which, as Davis notes, may even wink at the cosmic or sacred in the form of a neon Buddha or a garish Ganesh, as if to say, the times are such that we can only ever have our 'mystical experiences' within quotation marks. Throughout the Fifties and Sixties the drugs themselves were being manufactured by the Swiss pharmaceuticals giant Sandoz at practically no cost, and distributed through global supply chains to be prescribed/used for early forms of psychedelic therapy. As a result, the street price was low, permitting all-comers. This didn't stop certain more rapacious psychiatrists administering it to West Coast glitterati at the Sixties equivalent of today's psychedelic retreat centres for a hundred dollars a vial ($1,000 in today's currency). Aldous Huxley, whose name had become synonymous with the transcendentalising powers of mescaline, as described in his *Letters*, was by contrast appalled by the nascent LSD scene. 'I have seldom met people of lower sensitivity, and more vulgar mind.'[2]

For some reason, deep-rooted in our culture, we cannot help but see drugs as either sacred or profane. This extends even to the most 'innocent' of our habits. Think about the hallowed nature of the morning coffee or the evening aperitif. Is this revered anticipation materially different from that accorded to the day's first crack pipe? The psychonauts seem to be more alive to the puckish quality of distinctions between sacred and profane, sensing that even though LSD is iconically iconoclastic, there's a species of sanctity in the drug's godlessness.

Of all the psychedelics, Erik Davis told me, he believes LSD best embodies what he calls the 'tricksterish' world view. 'I took LSD when I was thirteen, so it's always been there. My whole intellectual and cultural sense of the world has always had an acid dimension. If your psychedelic experience extends back to adolescence, it's hard to think in scientific terms about how the drugs might alter one's "personal metaphysics"; there's no "before" or "after" conditions – it's all made of the same goo.'

Palmer has this quality, and Oberon has it too. And like Oberon, Davis, in opposition to 'scientism' (meaning the ideology of science), though in keeping with the purity of the scientific method itself (which is without ideology), has made a virtue of 'not knowing'. 'I think acid has deepened my relationship with uncertainty, that "not knowing" is as important as "knowing", and that how you practise your "not knowing" is key. There's a dumb version, the "I don't know, I'm just gonna do whatever I frickin' feel like" uncertainty, but that's not what I'm talking about. I mean that there's always a beyond, and any framework you have – whether it's religious or scientific – is not going to hold together in the face of the big, awesome, glittery void . . . That's what LSD has given me: not the conviction there is another world, or that the mind is eternal et cetera. But a sensibility of openness, a mischievous relationship with the sacred and the sense that the sacred is not just serious, that there's something furtive and tricksterish that's also in the picture.'

Like Davis, Palmer first took acid in his early teens, and in his own idiosyncratic way my new trip-sitter was keeping the psychonaut tradition alive. By conventional markers of expertise Palmer had 10,000 hours of tripping under his belt by the time he hit twenty, and his 'practice' had not fallen off much in the decades since. But in trickster fashion, he did not present as your typical acidhead, but was shaven-headed, gaunt, with the deathly aristocratic air of William Burroughs. But Palmer was also limber and effeminate, with an unlikely fondness for cashmere and Southeast Asian cookery programmes. Blanched like mushrooms that grow in secret shade, he preserved his white skin like someone determined to remove all trace of his Florida roots. While it was common for psychedelic enthusiasts, like those we were now surrounded by in southern Mexico, to display their allegiance in luminescent tie-dye, Palmer went the other way. He wore dark cotton trousers, each with more than one secret pocket stitched inside by his personal Nepali tailor, that made a uniform of guerrilla-type obscurity and paranoia. The signs of being deeply immersed in esoterica were also there – a necklace of skulls from the Bön tradition, the mudra he kept making with the second and fourth fingers of his left hand, the mantra that looped in the recesses of his mind, leaping on occasion to the surface as a Sanskrit whisper – but largely hidden from view. Not standing out was what he stood for. He kept everything that mattered: hard copies of McKenna's *Food of the Gods*, Garab Dorje's *The Three Words That Strike the Vital Point*, a Swiss Army knife and, most importantly, his variety of 'source material' – in bags padlocked within padlocked bags, at the bottom of which was his laptop, opened by an entire library of passwords (he had once built a 'cyber wall' strong enough to protect the Tibetan government from Chinese hackers). Palmer's whole life was encrypted. As a young man he'd fantasised, like his beloved LSD, about a career in military intelligence, but his father's luxury golf-course business slumped, the girl in the soft-top

Mercedes left him for a jock and the teachers who'd told him he was gifted changed their mind. By the time he turned twenty-one Palmer had left the idea of America, and life's mainstream tracks, for good.

There were things that drew him out of self-concealment, raised his temperature, provoking sudden spikes of dyspepsia. Black eyes blazing, he'd drop his usual, painfully courteous manner – the legacy of a father who once fronted golf resorts for multiple Major winner Arnold Palmer, after whom he was named – and begin to hold forth with subdued but unrelenting force. (Like the stammer of Dale Pendell or Terence McKenna's curiously serpentine syntax, Palmer's speech had neuroatypical features – phonological substitutions and malapropisms – which might point to something altered in the language centres of many hardened psychonauts. It's only a hypothesis; as yet there's been no formal research on this type of yogi.) One felt, at such moments, that there was a pressurised deep-water table in Palmer, looking for an opportunity to geyser; a story that was always there, waiting to be told. It just needed its moment.

And this is not a metaphor because there was such a story, as this chapter will relate. It was the story of Palmer's life in encrypted, password-protected form: everything that he believed, everything that went wrong, everything that might have been, condensed into a parable which was also a philosophy. Its name, in the spirit of inexpressibility, had been collapsed by him into three words (and one word into three capital letters) that both struck the vital point and defied understanding: 'Told Properly ZTH'. Palmer's memoir was also an homage to the ultimate psychedelic koan: the will to perfect the expression of the ineffable.

We had travelled south from Guadalajara to the small mountain town of San José del Pacifico. Since the 1950s and Gordon Wasson's love affair with María Sabina, these mountainous cloud forests had become a mecca for gringo mushroom-hungry seekers. Bob Dylan had been there. So had John Lennon. Buses from Oaxaca to the coast were time capsules, from which stepped young New Agers: loping, dreadlocked, manoeuvring past one another without acknowledging their existence. Zombies, really, armed with didgeridoos, pan pipes, ukuleles, but also, these days, with smartphones and Blue-tooth wraparounds – McKenna homilies on Audible – to help them track down the food of the gods. Palmer had grown up a raver, I had been a half-hearted punk; neither of us trusted hippies.

And yet 'hippies' no longer felt like the right word for the flat-Earthers, the pantheists, the anti-vaxers and eco-warriors, the *new* New Agers: anti-government, anti-science, some high-functioning, others less so, who had headed with us to the mushroom forests of southern Mexico. I had encountered similar types at the ayahuasca ceremony with Sphinx in Guadalajara, and I would meet them again and again in Central and South America. They weren't so different from the many spiritual junkies I'd met on the meditation circuit in Asia: seekers, ungrounded, the same ways of talking, the same beliefs, however disparate and contradictory, constellating around the sun of their tribalism. It was somewhat ironic that McKenna was so often their guru, given his ambivalence towards the movement: 'I'm very uncomfortable with my position in the New Age – so much of it is an affront to rationalism. "Why me?" I ask the mushroom. "Because you're not a believer," it tells me.'[3]

I asked Erik Davis about the ways this tribe differed from his generation of psychonauts. 'They want to replace the old broken world view that led to their alienation – political or economic systems, the exploitation of natural resources, etc. They think, "I gotta get out of this." So they glom onto another different world view that seems to offer something like relief: "The forest is speaking to you", or, "It's the end of the world, but a new age is dawning, etc.", or any one of a number of far-fetched conspiracies, and hang their hats on it . . . Well I'm not sure how valuable that is.'

For the older generation of psychonauts, psychedelics were more about loosening your attachment to any single world view, rather than binding yourself to new anti-establishment ones. 'The trouble with New Agers,' Davis continued, 'is they tend to cling. Then they don't look at alternate views, or problems, or contradictions; nor do they look at context, at history, science and politics. And without any reality principle to guide it, the whole thing can wind itself up into something really noxious and inappropriate . . . The openness I'm talking about allows you to fully have a mystical encounter with the spirit of the forest, and then go drink coffee and read the paper and worry about the outcome of the next senatorial election.'

San José del Pacifico was no more than a small scar of shops running between forested hills, while new Swiss-style chalet guesthouses mushroomed in clumps here and there. You could buy Andean-style ponchos 2,000 miles from source, and terrible coffee, even though some of the world's finest beans were on the doorstep (sold to roasters in London and Berlin at twice the volume these days, thanks to the additional yearly crop gifted by global warming). But the most striking feature was that everything that could take on the shape or ring of mushroom did: water towers, houses, chairs, mugs, tables, topiarised trees and bushes, or the names of cafés, restaurants and home stays. Abundance implied.

Except mushrooms were nowhere to be found; not in March, anyway. We had either mis-timed the season or climate change had meant the season itself had mis-timed. After several dead-end introductions, we met one lady wanting to sell us a jar for 'five people' for $70 US – about 6 grams of mushrooms preserved under several fluid ounces of honey, estimated Palmer. Not enough for him alone, never mind the four imaginary friends. To Palmer's eco-anarcho-mind, the mushrooms should be free. Irate in his own muted way, he unlocked his backpack and, from a small bag deep inside, produced two cotton buds, their bouffants grey and greasy rather than the normal snowy white. The buds had been dipped in liquid acid and transported from Kathmandu halfway across the world, including a heart-testing layover in Bahrain, which still carried the death penalty for certain substances, though, as Palmer noted, ear wax was not one of them.

One story of Palmer's storied life was how he was always on a quest for mushrooms that ended with LSD. It was a metaphor for something. Where mushrooms were organic and sacred, LSD was profane, synthetic, renegade. Palmer would sound fundamentalist about his preference for mushrooms, would swear blind the two experiences were as different as night and day, though I never found his distinctions reliable in the scientific sense. Acid was 'heavy', 'strong', lacked 'spontaneity', had a 'plastic feel'; mushrooms were 'wry', 'organic', cognitive 'enhancers', a turbocharge for parallel processing. And his preference was telling in itself, because in terms of his character (and my folk theory of psychedelic preference, which was only strengthened by my encounters with the ayahuas-cery Oberon and the toadish Sphinx), LSD's wonkiness, the difficulty fitting it into any framework, its anti-capitalism, its pro-fane obsession with the transcendental, its exile from American institutions, its exclusion from the current scene because it was too long-lasting, too arduous to contain, its extreme potency even at tiny doses, made Palmer its living embodiment. As such, it was nat-ural he would prefer 'the other' to himself.

Palmer wasn't sure about our dosage. 'I dipped them in the bottle *quickly*,' he told me, chewing the end of one bud, as though speed is its own contraceptive. LSD was measured in micrograms, with 100 being a standard trip. But you could go much higher. In Big Sur I had met a guy called Rob, a former chef at the Esalen Institute – once the epicentre of the New Age and now a Google hang-out slash New Age theme park – who reckoned he could cook a meal for the guests on a 'wash-out' of 10,000 'mikes' (micrograms), or drive the precipitous cliff-top bends of Highway 1, which can befud-dle even sober motorists. 'Lucky for me I was born double Gemini,' Rob told me. 'It means however fucked up three of us get, there's still a designated driver.'

Inspired by the experience of mushrooms in Yosemite, I had planned to do a decent hike through the spectacular forested hills engulfing the San José del Pacifico. I'd found an eight-mile walk, ranked 'Expert' in difficulty, to the nearby town of San Mateo, thinking we would forage mushrooms as we made our way. Now we were on acid, which is twice the length of a mushroom trip, it meant we should aim to make it to San Mateo and back.

Really, though, the hike was a decoy. I was determined to finally unlock Palmer's story; to have him tell me properly, in all its intricacy, the story of 'Told Properly ZTH': a story that was its own kind of journey, rated 'Expert' in difficulty. I had heard it from different vantage points, in different iterations on many occasions, but I couldn't tell you anything about it. It was somehow both too prolix and ineffable, linear and concentric, as though the Strava GPS route in Palmer's neural circuitry was too complex for anyone else to follow. Or rather, I had been too straight to follow it. This time I wasn't: the shaman needed to take medicine in order to really see his patient, and to make sure something was brought back this time I would record it on my traditional doctor's Dictaphone.

'Do we have to have that fucking thing on?' said Palmer.

I reasoned that recording was part of his therapy, a bridge back to normal living.

Palmer began, as he often would, with 'The Experiment at La Chorrera', in which the McKenna brothers had attempted to bond the psychedelic compound of harmine (a beta-carboline found in a variety of plant species) and mushrooms with their own DNA, through the use of a set of specific vocal techniques. (I've already got the story of one impossibly intractable experiment to tell, so I'll leave it to the reader to fill him/herself in on the details of that one.) They hypothesised that the experiment would give them access to the collective memory of the species, manifesting an updated philosopher's stone, which in their terms was a 'hyperdimensional union of spirit and matter'. Fifty years later, Palmer understood the experiment in his own computational terms: it was a 'hack into source code' that allowed the brothers to introduce 'a meme into the collective consciousness about a cataclysmic event horizon'.

'Which turned out to be wrong,' I clarified.

'Well, that depends on your time frame . . . Fuck: we've already been here.'

The forest meant that the signal on the Strava map was patchy.

'I think this "here"'s different,' I said, consulting my phone. 'Yes, I'm pretty sure that tree wasn't in the old "here" . . . Are you feeling it?'

I was. I felt as though I was giving birth to a species that was too large for my frame. It was a hot day; the acid was strong.

Palmer explained that at the highest level of cultivation the practitioner (for Palmer, Dzogchen and psychedelics were interchangeable) can access *rigpa*: ultimate reality. 'In that moment, you have a brief glimpse of omniscient view. You understand the expression of all duality as it projects from a point of unity. You directly understand the reason *why* reality is. My mission all these years has been to engender the experience and document it while in its midst.'

Such moments were fleeting and unpredictable, but when Palmer happened on them across the years, they had the feeling of continuity. Each time he rediscovered one he felt as though he was picking up where he had left off, like children sometimes do with dreams, chasing the same story, the same feelings, across successive nights. These experiences were so alluring that the juice of Palmer's life had become a quest to return to the same mythic territory, and then return back from it with something to show, a proof of concept. But, as noted in the last chapter, preserving something of such experiences is beyond difficult. The one time Palmer was able not to return empty-handed, to do more than just babble like a one-year-old, is the essence of the story of 'Told Properly ZTH'.

Somehow our walk had led us to a roundabout; the one patch of unforested territory in a whole Canada of trees. The sun beat down on us. We crossed a road where concrete-bearing trucks flew past and headed off down a track where stoned-looking pit bulls and old men with machetes looked straight through us. They had seen bad ideas like ours many times before.

If I had let him, Palmer would have gone on for hours in his usual un-shepherded, wilfully abstract, techno-theological register. But this time I was determined to get him to speak as concretely as the trucks that hurtled past us.

'OK, OK! You're such a literalist,' said Palmer, straining his usual politeness. 'At the time in question, I was working for this huge sex toy operation in Vegas. This was the Noughties, around 2007.'

The register had suddenly become less rarefied, less obviously

theological. 'I was in the black latex heart of the digital porn industry. They had me working in the company's marketing arm on early
Web tech. I didn't want this for myself. I wanted to add value to the
world's cultural programme, not get its rocks off. But my dad died.
There was no inheritance. I had no savings. I needed a job. I came to
Vegas, which effectively meant choosing between gambling and
sex . . . Anyway, my boss, the VP of Marketing, had a boy-toy husband, and it was his birthday—'

'You mean toy boy?' As mentioned, Palmer made occasional syn-
tactic reversals.

'No, we called him "boy-toy" after one of the products. The VP,
who was surprisingly wholesome and familyish, invited the whole
department to join their birthday celebration. As usual, I begged off.'

At the time Palmer was living on his own in a one-bed which had
a 'big-ass basement' with no windows. 'The basement was ergo-
nomically perfect for balls-out tripping.' There was a huge casting
couch, an altar of three turntables with multiple sets of speakers, a
video projector facing onto a white brick wall.

'My perfect night in was a "base camp" heroic dose of
mushrooms – 7 to 8 grams – and from there I'd make multiple sum-
mits with NN-DMT. I had three or four vapes loaded and ready to
go on a table, and a small molehill of extraction close at hand if I
needed more. But on the night in question I hadn't planned to be in.
I had no source material left. All I had in the apartment was thousands of strips of blotter acid that a colleague gave me in a trade.
Making the best of a bad job, I put on some music – a dub-step track
with *icaros* jacked over it – and looped the closing scenes of *Blueberry* onto the back wall.'

Blueberry, a psychonaut staple, is remarkable for its astonishingly
accurate pre-AI impressionistic trip renderings, in an otherwise
world-historically bad Western. Oberon told me the sight of it,
when he screened it for a Peruvian shaman, was enough to make
the indigenous old-timer fall out of his hammock.

'I put a few tabs onto each of my eyeballs,' Palmer continued,
'and lay back on the couch.'

He described how his mind quickly turned to the work night out: how they were one big happy family apart from him; how this exile reminded him of his childhood: his mum re-marrying when he was four, a new family imported into his home, Palmer slowly defecting to the tree house at the bottom of the garden, then retreating further still – first into science fiction, then into his own thirteen-year-old assays in neuro-chemistry.

'I remember thinking about how we were having such divergent experiences: them in some chichi bar, chatting, laughing, chinking drinks; me on my own on a couch literally up to my eyeballs in acid.

'Then, somehow, I drifted into a much, much deeper place. I guess you could say I had a full-blown visionary experience unlike anything I'd had before, in which I found myself in the middle of a two-way communication with some kind of entity.

'Four weeks before the night in question, the company had their annual convention at some glitzy hotel on the main Strip. The VP had booked the author of *Chicken Soup for the Soul* as the keynote speaker. [This fact checks out: reality can sometimes feel hallucinatory.] But now in my mind there's been a loop in time, which meant that the *Chicken Soup* author and our VP were on stage at the event with me in the audience. Only I had bi-located and time-travelled, meaning I am also here on my basement couch five weeks later. And somehow I have this new-found power bestowed on me, which means that I am able to tell *Chicken Soup* and the VP what to say at the convention. I am, effectively, giving the keynote address to the audience through these ventriloquists, telling the employees my vision: not just for the company, not just for the porn industry, but for the country as a whole – where America went wrong, how it might yet correct its course . . .'

'So what was the entity?'

'I was. Or it was using me . . . I was uncharacteristically eloquent, visionary: stuff I never knew I had in me, because it wasn't in me – it was this entity – but I fucking agreed with every word of it. I wanted to stand up and applaud my own speech!'

Entity encounters are common in the psychedelic literature.

There is wrestling over the status of their reality. Neuroscientists prefer to think of them as the emanations of unconscious processing. Some psychologists regard them as imaginative manifestations of a collective unconscious. Then there are more far-flung understandings like Palmer's or the McKenna brothers', which regard the encounters as authentic contact with another, *more real*, reality.

In theory there is an infinite array of possible forms the entities might take, but, with the occasional exception (McKenna's Fabergé eggs, machine elves and self-dribbling basketballs), these forms have often been mundanely conventional in their weirdness: green aliens, talking animals and the like (in the wake of the Internet even the elves and basketballs have become commonplace contagions in psychedelic space). In some instances, people claiming to have had such an encounter then demonstrate knowledge they couldn't possibly have had otherwise, presenting at least a superficial challenge to scientific explanation. But many of the 'messages' received from such entities tend to be prosaic and banal:

> *There will be a storm before the calm.*
> *I will die in my Sixties from something I can prevent.*
> *He was teaching me the rules/regulations of the NFL.*

Entity studies also suggest they tend toward the positive: only 1 to 5 per cent of those studied reported a negative message.

Personally, I had not had any such encounter. I remained sceptical regarding their status, preferring to think of them as somewhat rational gestures towards the irrational: symptoms of the need to creatively disengage from our cultural insistence on order and reasonableness (despite all the evidence that neither actually existed), to overthrow the tyranny of the left hemisphere. I was also a little jealous. Trip envy is a real thing.

'A voice inside me told me it was crucial I document the speech,' Palmer continued. 'So I jumped up from the couch and grabbed a pen and paper. The pen I found was a company-issue "mantra" pen, so-called because it voiced, via a mini speaker, self-help-style phrases to support the workforce. So, as I'm trying to transcribe the beautiful pearls of wisdom that are being laid before the convention by the entity, the pen is emitting phrases like, *I'm a good person, I see myself with wealth and happiness, I deserve success now*, and other self-interested platitudes. To my horror these icky phrases back-flowed into the speech of *Chicken Soup* (i.e. me, i.e. the entity), and fucked it all up . . . What had been the equivalent of the porn industry's Gettysburg Address, a nuanced deconstruction of an American utopianism, was looping back as repetitive schmaltz – our national dream in mantra form – as the speaker was controlled by my pen.'

'So the pen had become the entity?'

'Right. I changed pens so I could continue making a record of all the incredible things that were coming out of *Chicken Soup*'s mouth. Except as soon as I start to write with the new pen, the convention speech switches to a shockingly confessional mode. Now *Chicken Soup* is, to my horror, confessing on my behalf that, ever since I walked through the company door, I've been skimming in one way or another, hitting bongs on breaks, doing my own research on company time, calling in sick if I had something else planned . . . grassing me up on every minor felony or indiscretion I've ever thought of committing to the whole convention.'

Things got weirder still. The new pen lost its power over the speakers. *Chicken Soup* and the VP were back in control, and what they were saying is that they'd known about Palmer's anti-company behaviour all along. Worse, they'd planned it.

'The trip has turned into a full-blown paranoid shit-fest. The boss and the guru are telling everyone in the audience that they were videotaping me all the way, caught every grift, skive, skim on company film.

'Turns out the whole thing was an elaborate set-up to allow *Chicken Soup* and the VP to prove a philosophical concept which will form the basis of a new self-help manual and guaranteed bestseller: how even a no-mates stoner-loser type can still make it in this country. Everything is nurture, nothing is nature. The American

Dream is alive and well, providing you follow their step-by-step guide at $25 in hardback. As I continue to write down their speech, I have flashbacks to all my furtive misdemeanours, and instantly I'm convinced that they really were *letting* me misbehave as part of this terrible corporate social experiment, that *Chicken Soup*'s new best-seller will have my grinning skull blazed on the front cover.'

Meanwhile we appeared to have discovered a shortcut for our hike, a secret passage which ran along the side of the main road – a pavement, you might say, for that's what it was – that led us back to the same tree we'd been at an hour and a half before. Despite the dense foliage the sun had found a way to beat down on us through-out, reddening our mineral-free skin (we had lived on quesadillas for two weeks). Sweat made sight difficult. Water had become a problem. Progress was slow. Birds evolved overhead.

Despite these physical hardships, this time round I had success-fully tuned into every word of Palmer's story. It was no wonder I had previously found it impossible to follow: there were all these layers of cause and effect, these conspiratorial, paranoid shifts between subject and object, so many secret pockets stitched into the fabric of his thought. The story didn't seem to be under any teller's control, or else it was under LSD's control, a trickster intel-ligence that could take the significance of something and then morph it into its opposite – one moment liberal and anti-establishment, the next conservative and capitalistic – or move outwards to a larger frame, placing everything inside within quota-tion marks, again and again. It required extreme mental looseness in the audience, a looseness which I had, in mind and bowel. The net effect was to make the story like the day itself: a molten caul-dron: paranoid, hilarious, tedious, random, and yet inevitable, capable of going in any direction, always tending towards one, the one where you were always lost, the one you most wanted and feared, the one where you inevitably ended up falling in.

The science literature has tended to avoid the psychotic dimensions of psychedelic experiences, to emphasise the benefits rather than the risks to mental health. The majority of the general public don't see it the same way, probably harbouring similar fears to those that led to the banning of LSD and other psychedelic compounds in 1970. Since that time, through successive iterations of the *DSM*, psychiatry has attempted to specify the biological underpinnings of psychotic diagnoses, refining their reliability and validity as a particular pathology or disease. To date this has been spectacularly unsuccessful, as is currently reflected in the ad hoc and largely ineffective state of standard treatment for severe and enduring mental illness.

There have always been alternative, non-psychiatric ways of seeing psychosis, even within the behavioural sciences. The psychologist Richard Bentall has spent a career showing that the iconic psychotic symptom – hearing voices – is better thought of as a signal error, the 'misattribution of internal thought' to an external source, which can easily be induced in healthy volunteers under certain 'stressful' experimental conditions.[4] Other disciplines, from the social sciences to the arts, have advocated a different, more curious way of thinking of and relating to the voices – as poetic muses, as philosophical instructors, as delinquent superegos. These other approaches have become a buttress against the medical model, giving rise to self-help groups like the Hearing Voices Network.

It was from this broader clinical perspective that I attempted to understand Palmer. While his story sounded like psychosis or a delusional syndrome, it was – we were – under the spell of LSD. Clearly it had left its mark on his sober life, which was eccentric if not dysfunctional. Any formulation would also have to contend with the apparent trauma of early abandonment, and his Asperger-like way of seeing the world. But then there was the world itself: America, Vegas, the porn industry, self-help, exploitation, aspirationalism. The signals were so diffuse, so confused – how could they not be misattributed? Palmer had found a way, healthy or otherwise, of protecting himself, of finding security. His behaviour may have been risky, but to him it was safer than the thing he feared. It placed him in the tradition of psychonauts, experimenters who were choosing, more or less consciously, to allow their grip on reality to loosen, to dip their toes in latent psychosis, either through intensive first-person research or, in Philip K. Dick's case, a mixture of prescription drugs and his own imagination. If there was something maladaptive about it, then it was more of the order of the 'clinging' Erik Davis had spoken of with respect to New Agers: that Palmer held onto his story too tightly, had let reason drift too far in the rear-view mirror, stopped reading the papers covering senatorial elections.

'The punchline of *Chicken Soup*'s joke and the climax of his new book was the announcement that I was going to be the porn giant's next CEO. Triumph or humiliation, depending on who's looking. *Chicken Soup* and the VP are calling me up on stage at the convention, but I'm not there, because I'm five weeks in the future on my couch, humiliated, terrified, and somehow concerned I'll be late . . . And just like in dreams, the whole trip unravelled as my name was called out, the audience looking around, waiting for Arnold Palmer McMasters to leap up on stage and take a bow . . .'

Palmer's story balanced narcissism and paranoia perfectly, and showed how one could easily tip into the other, even that they were made of identical material. I'd lost all reception on my phone a long time ago, so as we hiked through painfully familiar forest without a map, Palmer continued describing how he had written everything down on a notepad in tiny writing: all the wisdom, the thought insertion, the mantras, the corporate bet, the humiliation, the climactic promotion to CEO. Sobering up, he had changed the music on his decks to chill out, eaten some food, smoked a joint and drifted off to sleep.

'Now I was the only person in the house that night. The door was locked, there were no windows in the basement. I made sure the notepad was safely on the table next to me as I slept. But when I woke up a few hours later the notepad had disappeared, gone, vanished into thin air. I have never been able to find it.'

As with all professional paranoiacs, Palmer was famously meticulous about the security of his possessions.

'How do you explain that?' I asked.

'You clinicians with your leading questions.'

'I'm just being curious.'

'You look fucking curious too.' I made the cardinal psychedelic mistake of looking in the (iPhone) mirror: my face was a pan-fried quesadilla with dark green lips. 'I don't know, man. It's hard for me not to get conspiracist. Maybe, just maybe, the notebook had been somehow called back by the entity to the reality where it normally hangs out. Or, as you and your colleagues would say, it not being there was documentary proof I had erased my mental health.'

But I didn't think that; not straightforwardly, anyway. Neither would Erik Davis, whose book *High Weirdness* is fascinated by occult thinkers wanting to understand the fantastic as somehow being part of natural law, a will that allows them to bend all rational frameworks to it, fixating on anomalies as proof of a deeper, more essential order that is somehow more real.

And outside of the story, it felt as though the LSD had somehow created the perfect frequency for Palmer to be at his most coherent, and for me to be at my most attentive. Our antennae, to use Hofmann's metaphor, had fused with one another and the world around us. By this point in our journey it just felt more prudent to trust the tale, not the teller. Or rather to trust the tale's mortal hold over the teller (if Palmer was the Ancient Mariner with his burning eyes, then he was in the grip of a still more ancient, more intense mariner), a tale so magical it had its own disappearing act.

'I was devastated, heart-broken, like I'd lost proof of this most precious thing,' Palmer continued. 'That was a Friday night. The next day I did some yoga, meditated, ate some salad. I told myself, *I will never take drugs again*, over and over . . . By Sunday, I was like, "What the fuck was all that about?" By Wednesday I was wondering if the same outcome would happen again if I repeated the experiment. Yeah, I know, the definition of insanity and all that crap . . . On Thursday, it was, "Do you realise what you have done? You have found a hack on ultimate reality: your life's work. You owe it to yourself to give it one more try."'

Friday, one week later, Palmer took exactly the same dosage of LSD in the same way – same music, same video, same position lying on the couch – like the infant trying to catch the tail of last week's dream. This time, however, he had his laptop set up on a table ready to digitally preserve whatever the entity had to say.

'It was all straightforward, formulaic even. Within half an hour my entity is waiting at the bottom of the rabbit hole. Only it's so much more straightforward this time round: no mutant mantra pens, no *Chicken Soup* paranoia. I was able to ask the entity in a relaxed, conversational way about whatever I wanted – as though we were sat in a quiet booth by the open fire of a British pub; able to dive carelessly into all the mysteries I've spent more than half my life obsessing about, only to have the entity decipher and explain them with astonishingly simple lucid answers . . . That was the thing: everything was so self-evident, so simple, so mundane, so *straight* . . . And this time round I was able to transcribe every word on my laptop. He'd even repeat bits slowly if I missed them first time round.'

This felt like an island of sanity within the clinical picture: the relief of the ordinary amid bizarre inter-dimensional communication. It brought to mind the British paediatrician and analyst Donald Winnicott's summary of his life: 'All my life I have been imprisoned, frustrated, dogged by common sense, memories, desires and – greatest bugbear of all – understanding and being understood.'[5]

What Palmer woke up to find the next morning was a transcript that perfectly reproduced his conversations with the entity. But of course, as the world's experience of ultimate reality did not undergo some sudden seismic revolt some time in 2007 (at least not one we are aware of), there has to be a twist to this story. And here it was. While Palmer's questions had been transcribed word perfectly, the responses of the entity, those pearls of extraordinary insight, had been transcribed by the letters 'OMFG', over and over and over again. I never actually saw the original Word document, but Palmer described his dialogue with the entity verbatim in a tone of heart-broken bewilderment.

'. . . That's truly fascinating, thank you, sir. And can you tell me one more time, the key to the philosopher's stone?'

'*OMFGOMFGOMFG OMFGOMFGOMFG OMFGOMFGOMFG.*'

'*Really?* It's that simple? . . . Wow. And would you just run past me your solution to the primordial riddle of consciousness?'

'*OMFGOMFGOMFG OMFGOMFGOMFG OMFGOMFGOMFG OMFGOMFGOMFG.*'

'Amazing! That makes total sense. Thanks again . . . One more thing. Would you mind telling me if anything David Icke has said is actually true?'

'*OMFG OMFGOMFGOMFG OMFGOMFGOMFG*', etc.

'*Incredible* . . . I will spend the rest of my life trying to explain all this.'

I remember how thwarted and humiliated Palmer looked under the shade of our favourite tree, a giant fir, under which we had finally come to rest.

'There was one thing I managed to bring back – one bit of dialogue where I had incorporated a bit of the entity's direct speech into one of my clarifying questions. This is the bit where I say, "You mean to tell me, "*Told Properly ZTH*"? That's, "Told *Properly ZTH?*" . . . Got it, thank you so much.'

'What's ZTH?' I asked.

'In computer language "ZTH" is a variable holding the value of infinity. The implication being, that if infinity was communicated properly – as the entity had communicated it to me in such simple lucid terms – then it changes everything. *Ev-er-y-thing.*'

And because it couldn't be, the same 'everything', including and especially Palmer's life, remains unchanged. A parable about the quest for ultimate reality, which is also, inevitably, a parable about deception, self-sabotage, insanity, stuck-ness; believing you've made a lifelong journey, when really you've never left your childhood home, or that tree very near the start of your expert hike.

Unless, that is, you're a professionally defiant monomaniac in the style of McKenna; in which case it's a parable of the necessary ineffability of all interdimensional conversation, or more urgently, that our understanding of the laws of ordinary reality is profoundly inadequate, in need of alien help.

I chose to look at Palmer's 'Told Properly ZTH' in a different way. Not as a clinical case study. Not as a test case in the nature of entity communication. But as a work of fiction.

It's a commonplace to claim that psychedelics make you creative. During the first wave of psychedelic research in the Fifties and Sixties this became a major region of interest. LSD, in particular, was a hypothetical enhancer of creative thinking, which either focused attention and/or broadened the attentional scope. It was seen as permitting dynamic states that allowed moving between different modes of thought that generated insights, as demonstrated by a host of paper-and pencil-based experiments that ran alongside a library of anecdotal reports. Artists and other 'creatives' (the moniker itself is LSD inspired) used the drug to enrich output and add novelty. This stretched to claims of their centrality in major scientific breakthroughs such as Kary Mullis's Nobel prize-winning discovery of the polymerase chain reaction.

The current wave of research has continued this project in more piecemeal fashion, adding million-dollar neuroimaging suites to the pencils and paper, though tending to switch out LSD for more medically fashionable psychedelic drugs. For instance, according to the previously mentioned anarchic brain hypothesis, psychedelic creativity is the ability to shift between the generation of novel ideas and their evaluation. Under psychedelic influence one's 'system' of thought is temporarily unconstrained, allowing greater flexibility to imagine different possibilities. Thinking becomes hyper-associative, including 'contradictory or illogical thoughts and feelings, the transformation and merging of images, and illogical and abrupt transitions between thoughts'.[6] The researchers note that creativity can also feature 'compromised reality-testing', which makes it vulnerable to 'magical/wishful fantasy-based thinking'.[7] The psychedelic mode of creativity is not exclusive to drug-induced states, but can be found in other altered states of consciousness: dreaming, trauma, trance, holotropic breathing, sensory deprivation and, of course, psychosis.

Palmer's research had not taken place in a laboratory. As with McKenna's Experiment in La Chorrera, Palmer's spectacular failure according to the terms of science was, to my mind at least, more than adequately compensated for by its artistic merit. However Palmer saw it, I regarded 'Told Properly ZTH' as a work of literary metafiction, part of an LSD-inspired genre of narrative loopiness and paranoia that included Thomas Pynchon and Philip K. Dick and had helped shape contemporary storytelling culture. Whatever adaptive or maladaptive function it served in his personal psychology, it was also an expression that belonged to Palmer rather than an entity; only he could have told 'Told Properly ZTH' properly. And yet in terms of the anarchic brain, it was a story that, therapeutically at least, had trapped or *canalised* his life in a narrow, uncreative space, as much as it had permitted unconstrained freedoms. 'Once you get the message,' Dennis McKenna wisely said after his psychotic break at La Chorrera, 'hang up the phone.' Instead, Palmer had kept on the call and placed the rest of his life on hold.

Under our fir tree Palmer continued to squeeze out the last of his devastation. 'I fell on the floor in paralysis, holding my stomach, literally gutted. The "why" of reality remained just out of reach, hopelessly enigmatic. I felt transcendence edging me like a veteran porn star . . . Something snapped inside.

'I went to my boss, the VP, the following morning. I made him an offer: together he and I could revolutionise the entire sex shit-show. I set out a plan, linking our entire marketing output to newly emerging social media platforms – this was 2007; they had only just started, and people weren't that interested. I told him that the right well-placed memes would spread like an STD all over the Internet, without us having to do anything. It really *was* the future.'

There *was* something visionary about the net effect of Palmer's LSD journey after all. Only it was more mundane and practical.

World-changing certainly, but less esoteric, less spiritual, than he'd been looking for.

'Meanwhile, I was dimly aware that I hadn't changed my clothes all weekend, and that I was speaking without breathing at an unbelievable velocity. I could sense the VP was getting over-whelmed, but I just couldn't stop the words coming out – more and more of them, it was definitely a kind of mania – detailing how we would with the right suite of strategies become market leaders, pion-eers of change, create a new breed of product . . .

'Eventually the VP says, "Palmer, Palmer, Palmer – these are great ideas, really they are. But, listen: we sold over a hundred million dollars' worth of sex toys in 2007 doing things the old way. You're a bright guy, Palmer, a talented guy. I believe in you. You have a knack of looking round the next corner. So why don't we try out some of these ideas over the course of the coming months in a staged way? That makes good sense to me. What doesn't make sense is total revolution; to turn this ship on a dime on a stoner's hunch when we are making so much fucking money. That would be, would be" – I still remember the way he was looking at me – "fucking insane."'

When I came months later to transcribe our conversation there was a pause at this point on the recording for some seconds before Palmer's voice began again.

' "You're absolutely right," I said. And I quit there and then. I had no idea I would quit before the words came out, like I'd swallowed the talking pen. In the space of a week, I had gone from being CEO to unemployed . . . Well, that was just about it for me and capital-ism, the end of my walk-on in the American Dream. Within a month I was on a plane to Iquitos and the Peruvian jungle. That was a decade ago, and I haven't returned to my country of origin since.' Palmer looked at me as though waiting for the diagnostic axe to finally fall. 'Listen, I'm painfully aware of how all this must sound to someone of your training. But I would say the grandiosity, the messianism, is somehow just riveted onto LSD. Really it's more of a John the Baptist than a Christ slash Neo complex. I mean, look at me. Do you really think that I think I'm the One?'

I could feel tears rising, but I didn't know if they were mine or his.

'There's nothing special about me – I could be anyone. What's special is the message itself, which has the potential to disarm us of all illusion, free us from clinging, re-boot the whole fucking planet.'

Creative though it certainly was, it was also clear, tragically clear, that fifteen years later he still lacked authorial distance.

Then, from nowhere, a new surge in the mains of his circuitry. 'Don't you see? It has to be told properly. *ZTH has to be told properly*. It's ineffable, but with the right blend of artistry, imagination, a psychedelic sense of connection, a mash-up, of humour, high feeling, rigour . . .' He was burning holes in me with his blazing eyes. 'In so many ways I've left my earthly life behind trying to bring this monster back, trying to tell ZTH properly, but I just can't do it . . .' Still he keeps looking at me. 'And then I met you, in a Himalayan café, and you told me you had the idea of writing this book, and instantly the thought came over me: "It's him. He's the one. His book will be ZTH *told properly*." '

The author is always the last to know. Finally the penny dropped. Creativity's last loop. This wraith-like man, whom I had assumed I was recruiting to be my trip sitter, has spent over a year believing that *he* is recruiting *me*, as an amanuensis: to write the mystery of ultimate reality, to tell the story of ZTH properly, as I have just done, to the best of my abilities.

I start to laugh; the laughter of the lifelong seeker finally finding his jewel in the mud, which is also the laughter of the same man watching the same jewel sinking back into the swamp. In other words, I have been possessed by Palmer, both of us laughing harder and harder. I'm high enough to have the thought that I'm laughing Palmer's laugh rather than my own, while he's got hold of mine. And without telling him my thought, I'm convinced he's thinking it too. The laughter has become clinical in severity. It feels as though we're edging into non-dual realms.

Only, and many minutes later, after the laughter has finally been tamed, do we really take our bearings. Palmer and I find ourselves sitting in the middle of the same traffic island that siphoned the flow of concrete trucks we had encountered several hours before, near the beginning of the journey. The Strava map indicates that, rather than an Expert-level hike to San Mateo, we have spent the last four hours circumambulating this holy site, literally walking round a circular island in circles, without knowing it. We laugh some more. On my phone the digital profile of our walk looks like the rings on the cut trunk of a tree, or like a man wearing a watch milometer who has stroked his chin over and over at a bemusing thought, or has pleasured himself fruitlessly for several hours – both of which, we agree, is what we have effectively been doing for most of the day. Then we laugh the laugh of the still more mad, marooned on an elevated traffic island overlooking the wonderful un-trespassed forests of the San José del Pacifico valley. We upload our walk onto Strava, label it 'Expert' level, and joke how this is the only meme Palmer has brought back from the entity, and how, unlikely though it may seem, its material reality might just change the world.

If we had taken mushrooms that would be the end of the story. As it was LSD, we still had four more hours to endure.

We decided to give up the illusion of a hike and make our beds on the traffic island. Each of us plugged in noise-cancelling head-phones (which only seemed to concentrate the noise of the traffic) and listened to some music. Inspired by the thought that a tripping John Lennon might once have passed by this exact spot, we made our way through the entire catalogue of the Beatles.

Neither of us had listened to them since our teens, but as they travelled from working-class, psychedelically naïve Liverpudlian geniuses to stoned, world-bestriding geniuses, it became possible, we discovered, to trace the development of their creativity – the shifts in musicality, the gradations of lyrical strangeness, the incremental weird-ifying of their world – as the expression of the band's experimentation with LSD, from the first murmurs of it on the perfect pop of *Rubber Soul* in 1965 to its apogee on 1967's *Sergeant Pepper*. Our playlist arrived at the album *Let It Be*, produced in 1970, the year the Beatles split up; the same year that psychedelic research was banned in the US.

'That's after "Strawberry Fields"?' said Palmer.

'Four years after?'

'Why a strawberry?'

'It's the name of a park Lennon played in as a kid. But the meaning doesn't matter. It's just the perfect dog whistle for those that *know*.'

We listened to the music for a while. I thought about how my research kept getting hijacked by experiences other than my own, that my journey was always following someone else's map. And what was I really learning? Was I unearthing something profound about psychedelics themselves or was I just adding to my clinical load? Was the current escapade really about the visionary qualities of LSD; or was it about Palmer, a McKenna manqué, elaborately self-involved, creatively paranoid, hysterical even, and, like his hero, a gifted raconteur who adds but little to the sum of human knowledge.

It's my turn to tell Palmer a story. Two months previously I had watched Peter Jackson's *Get Back*, a documentary about the recording of *Let It Be* over twenty-two days in January 1970. In my opinion this film is by far the most creative and interesting of Peter Jackson's various multi-hour epics. What you get back from *Get Back* is almost everything: the most astounding moments of creativity – McCartney's fingers stumbling in real time on the title song, which at the level of synaptogenesis is the same as watching new dendrites sprout on an MRI, only more rewarding – but also lassitude, bickering, child's play, over-excitement and sulking. A slow-burn soap opera: three weeks of studio time to record the whole album, dramatising the different ways four can be split and split again, as the music keeps on coming, against a calendar of passing days.

It builds to that iconic moment when the group give their last ever live performance, and the first in over three years, from the roof of Apple Studios. Meanwhile the police, alerted by multiple complaints about noise, are on the ground floor negotiating with brilliantly evasive door staff about the need to keep it down. And on the streets outside the studio building, a news crew interview various incarnations of 'man in the street': men and women of every warp and weft, every age and social class, voxing their view of the band amid a split screen of live music and police negotiation. A couple of naysayers aside, who only serve to highlight the prevailing feeling of innocent, heartfelt appreciation, their message is the unspoken message of the whole film. Which, if you will allow me to extrapolate a little, is this.

These few days in January 1970 are really an inflection point in world history. This is the last time that the best, most creative music in the world is also the most popular. In a few months the band will split up. Acid will be banned at roughly the same time. Over the coming years music will switch from analogue to digital. Tastes will fragment in compelling ways. The man in the street will get used to being filmed – will start filming himself ad nauseam, giving out unrequired opinions that are polar and polarising. Soon there will be talking mantra pens, *Chicken Soup for the Soul*, a multi-billion-dollar porn industry, several mental health epidemics, one Psychedelic Renaissance and counting, and the promise, to borrow from Sam Harris, of the first-ever fist fight on Mars.

And there in the middle of that moment are four men who by this point more or less hate each other, men still in their twenties who, for twenty-two brief days, despite or because of years of psychedelics, were able to summon all their maturity and creativity and flexibility to work something through, like the last adults in any room. Men who knew that strangeness, humour, self-deception, tricksterishness could reconfigure the tastes of the whole world as long as they were more or less under the control of extraordinarily musical minds.

'Which is as if to say,' I told Palmer, my turn to do the button-holing, 'that's it. That was the end of the golden age. We had our chance, and we blew it. Not just the Fab Four. You, me. All of us. Psychedelics were banned a few months later. The culture moved on, *"progressed"*, when really it began to slide. From now on, everything that feels like a renaissance will really be a little bump on the downward slope, another circuit round a traffic island.'

But, honestly, I didn't need LSD to see it that way. The next morning it was time to head to the real south, to one of the few places on earth that had never heard of the Beatles.

PART THREE

8

'You Will Have Another Child'

The Kogi homes were simple wooden frames. Two were completely open to the mountains, the other had mud walls and each had steep-sloped roofs of plaited banana leaves. Around them were small crops of corn, cotton, coca leaves, a pile of yucca and a small balding black pig on a leash. Naked children were taunting a scrappy-looking dog. An older girl was building a bamboo fence with a machete. One of our group, Santiago the artist, greeted an elderly lady, the aunt of Daniel the moma, who was sewing in her hammock. She handed him the *mochila* she'd woven to order. Santi would make a cast of the small bag and use it to sculpt an 'indigenously inspired' piece that had been commissioned by a luxury hotel in the city of Cali, hundreds of miles to the south, for its refurbished atrium. He paid her with pesos and handed over a gift of beef steak and cheese from the local store. The Kogi don't use money among themselves, but need it when they do business with 'Little Brother'.

After some black tea Daniel led us out of the village towards the snow-capped mountains of the Sierra Nevada. He walked gracefully, I thought, no more than the size of an average thirteen-year-old boy in the West, with shoulder-length black hair. He was in his mid-forties, I reckoned, but I might have been as much as twenty years out in either direction. He was dressed traditionally: a white hat in the shape of a Hershey kiss, a self-tailored white smock that fell to just above his knees and boot-cut white trousers. Later Santi told us Daniel was not his name, not even a Westernisation of his name: it was an avatar of an avatar – his real identity buried out of sight. But I already knew that what I saw wasn't the thing itself. After weeks of psychedelic research I was more aware than normal that my set meant I couldn't help but see everything through the clichés of Western eyes, unconsciously layering what was there with unknown assumptions, like seeing Daniel's boots as Wellingtons. Until Santi explained they *were* Wellingtons, their trademark squeak an alien note in an otherwise natural soundtrack. Really it was we who were the aliens here.

We arrived at a clearing in the brush, and Daniel sat himself down in a cradle made by the roots of a mango tree, his back resting on the trunk. By his feet several stones were half-buried in the ground at crooked angles. This was evidently a sacred site, but that had only become apparent now I was in the middle of it, paying attention. From the outside – incurious, unaware – the meaning was undisclosed.

Daniel took a bottle of water from his *mochila* and filled half a coconut shell that rested by his side until the water brimmed at the top. He took two green tubular stones, *tuma*, small as wisdom teeth, from inside his smock. Each *tuma* had a small hole bored through its centre. He dropped them one by one into the water. As they sank, water was forced through the holes, and a flurry of bubbles rose to the surface. Daniel tracked the bubbles. According to the research of the anthropologist Arbeláez Albornoz, the size, number, duration and movements of the emerging bubbles form multiple codes, readable as the meaningful replies of water spirits made through the *tuma*'s 'mouth'.[1]

Daniel scrutinised the earth around the shell. He looked at the moving leaves of the palm trees above his head. He took some coca leaves from his *mochila*, placing some in his mouth, crumbling the rest onto the surface of the water. He always kept a wad of leaves packed between his gum and cheek, which gave the side of his mouth a *Godfather*-like tumescence. In one hand he held a large orange-yellow gourd (the *poporo*) which had lime at the bottom, made by baking and crushing seashells collected from the coast. With the other he sank a moistened wooden needle into the gourd repeatedly until it was coated with white powder. He inserted the needle into the corners of his mouth, activating the psychoactive properties of the leaf. He spat out some saliva and stared into his own spit as if it was a night sky.

After a moment he spoke to Santi in broken Spanish.

Santi translated.

'He says it's good, the divination can begin.'

People tend to behave as though where they live is the centre of the world, but for the Kogi Colombia's Sierra Nevada is literally the heart, the Mother, the nucleus of the cosmos, the place of the vortex. The landscape, a mountainous island disconnected from the Andean spine that rises straight from the ocean to 6,000 metres, 150 miles north-east of where Panama joins the South American continent, seems to be on the Kogi's side. We had hiked up here earlier that morning – Santi, Palmer, myself and an Eastern European man called Dmitri – from the beach town of Palomino on the Atlantic coast, for a divination ceremony using coca leaves, led by Daniel the moma, a Kogi priest.

The Kogi are a tribe who have inhabited this region for over 1,500 years, protected from the different waves of colonialism by the natural sanctuary of the mountains of the Sierra Nevada. To the Kogi, living there meant tending the Mother on behalf of 'Little Brother', their term for the rest of us. I'd been connected to them by my friend Ben (Izza's father, whose kitchen had been the site of my ketamine experiment, which, though it felt like years, had been less than six weeks ago), who had helped to make the feature-length documentary *Aluna*, a Kogi term for 'mother', denoting both cosmic consciousness and the mind inside nature. The film was inspired by their desire to communicate to the wider world the ecological disaster we all faced, as they discovered it in the melting ice caps, the shifting seasons and the disappearing flora and fauna in their heartland.

Although I found *Aluna* largely impenetrable, it conveyed how in Kogi cosmology the mountains of the Sierra were microcosmic: everything that happened there was related to everything beyond it

on a scale that radiated out to the entire universe. The delicate balance of the whole was maintained by making 'payments' to the
Mother: symbolic offerings of food, dance, song, bodily fluids, special stones carried from other parts of the Sierra, which bound
them into a constant expression of gratitude and reciprocity for
the ineffably complex web that kept them alive, a breath at a time.
There were microcosms within the microcosm, a sort of fractal
ecology, so that changes in the region could be discerned in the
minute shifts in the Kogi's own bodies. The body itself, meanwhile,
was also a map of the universe or, up close – supported by the
coca – of the particular quivering flight of a butterfly, or the changing temperaments of the wind. (It reminded me of Merleau-Ponty:
that objects had a secret life of their own, that 'experiencers' –
people – must attain the right 'body set' to share in it.) To this
almost impossibly sensitive way of seeing, any single thing was
charged with everything, the minute with the macro, the physical
with the metaphysical, the present with the past and future: all was
inter-related, overlapping, stitched into itself with the fragile thread
of life. And sewing, normally the preserve of men, not women, in
Kogi culture, was deeply woven into a cosmology that regarded
the spirit of the Earth, known as the Mother, as a textile. The intention that lay behind every action was a thread stitched into the
fabric of reality. All creation, all behaviour, all thought, was *sewn*.

Textile, but also text: Daniel lifted his naked foot from the sandy
ground and read the print it left. He raised his head to the skies and
listened to a songbird. He looked at the different veins of leaf-cutter
ants as they transported their goods from the base of the tree. He
put more coca in his mouth, spat once again and closed his eyes.

For the Kogi, coca leaves give mental clarity, which promotes the
sense of connectedness to others and the Mother necessary for divination rites. (In the terminology of Western neuroscience we
would say the leaves enhance 'focused attention' and 'social cognition'.) They are also stimulants to aid with farm work or for walking
the prodigious distances needed to constantly caretake the whole
territory. Coca's mildly euphoric effect elicits dancing and music,

which are thought of as social forms of 'payment'. It also helps with sexual arousal for conception. Coca both dampens hunger and provides nutrition, being rich in vitamins, which is important in the context of the relative scarcity of vegetables in the Sierra. It is also a symbol of reciprocity: whenever one Kogi man encounters another, leaves are swapped as a mark of respect.

All of which is to say that reducing coca to simple, mechanistic 'drug-effects' – the terms of reference of Western science – is to literalise and under-sell its multi-faceted meaning as a node linking together different facets of Kogi society and cosmology. Meanwhile, less than three miles away, we had spent the previous evening watching a group of young men and women from London cutting lines of a chemically treated version of the same leaves (of scarcely credible purity, for which they paid $5 a gram), disconnecting themselves from everything about the strange reality they found themselves in, preferring the loud, shrill echo chambers of their own voices.

Daniel looked down at the water's surface as it lapped against the coconut shell. He dropped the small stones in once more and read the bubbles. He turned to Santi and spoke.

'He says we have brought dark energy with us to this place. We must clean ourselves before we can go on.'

Because each one of us – Santi, Dmitri, Palmer, myself – was neurobiologically conditioned to believe he was the centre of the world, each of us also believed he was the main culprit; the amalgam of grandiosity and self-deprecation being a particularly Western set. Daniel took out a lime from his *mochila* and cut it into four pieces. He looked at the way the cells of the fruit were arranged in each segment before handing one to each of us, telling us through Santi to draw out the 'bad juice' from inside. As we sucked on the lime Daniel got to his feet and picked some cotton from a nearby bush, passing us each a small ball. We should spit all of our badness into the cotton, focusing the dark material – misdeeds, past traumas, long forgotten lies – into its whiteness as we spun around and around on the spot. This ritual was our payment to the Mother; now that we were clean, our past and future lives could be seen. He collected our balls of cotton and told us he would bury them in a specific place later. Then he placed some more leaves in the corner of his mouth and watched the new world unfold before him.

I wondered what it might be like to see as Daniel saw. His formation as a moma, orders of magnitude more extreme than any monk, required him to spend the first eighteen years of his life in a darkened cave. Those first years when the brain is at its most plastic, neurons arborising and pruning according to inputs received, include multiple critical windows of development, which if missed can never be re-opened. That, at least, is the scientific wisdom. Yet Daniel gave the impression of seeing everything. (Though Santi had heard rumours that the young momas came out at night and saw the world by moonlight.) The cave was really a hut, or *nuhué*, the design of which re-created the nine vertical levels of the Kogi cosmos: upper worlds of physicality and light, lower realms of spirit and darkness. Within this there were halves and then quadrants reflecting different dualities: sun and moon, sea and summit, body and mind and so on. The young momas would be trained by elder momas in all the aspects and practices of Kogi spirituality and cosmology, of which divination – which was not so much fortune-telling as reading the world as it manifests itself – was central. The momas had never been near a neuroimaging suite, but I imagined their neural architecture would be as distinctive, albeit in subtler ways, as their cosmology and metaphysics. What would seeing be like if that architecture had been so exquisitely trained in noticing everything *from the inside*; in approaching reality without the normal priors but as something always renewing itself, unstable, unfolding, precious? That's what I found myself trying to imagine, when really it was inconceivable.

Daniel opened his eyes, dropped the *tuma* into the water once more and watched the world unfold.

Erik Davis had told me that all psychedelic roads lead south, sooner or later. In recent years Western research and business had begun to make the journey, negotiating with the political, financial, ethical and metaphysical implications arising from the clinical and commercial instrumentalisation of indigenous plant-based practices. This was either an authentic sign of maturation or a superficial gesture of virtue signalling, depending on one's perspective. What was a fact was that less than 1 per cent of the hundreds of millions spent on developing and capitalising plant-based practices had accrued to indigenous peoples. 'Psychedelic Renaissance', a recent paper,[2] celebrated the ironic appropriateness of the name; like its sixteenth-century European precursor, the whole enterprise had been built on the foundations of 'colonial extractivism'.

The Kogi did not have a known tradition of classical psychedelic use. Santi told me he had heard that some of the elder momas in the High Sierras used 'special mushrooms' for medicine, meditation and dancing. But this was never talked about openly. I had been drawn here because there seemed to me to be something so inherently psychedelic, so world-manifesting, about Kogi culture itself. So far as I could grasp it, their understanding of reality, though rudimentary from the perspective of Western science and naturalist philosophy, was intricately multi-dimensional and self-reflexive. The different tools and concepts science had used to investigate the psychedelics – music playlists, nature-relatedness, mystical experience, connectedness – appeared almost comically rudimentary next to their rich elaboration in Kogi culture. What might a Kogi person make of ideas like 'set' and 'setting', I thought? How would it be possible to define or constrain them in the terms of Western science, when anything that had ever happened to anyone was also happening to everyone at every imaginable time and place? How to 'operationalise' something like 'nature-relatedness' when there was nothing in the world, concrete or spiritual, that was not related to it in dynamic, multidimensional ways? What could a 'mystical experience' mean if every experience contained the same mysterious seeds? How could something be noetic, inexpressible, revelatory, if reality was always expressing and revealing itself in every moment in a way that was by Western standards inconceivably heightened?

There was an analogy to be drawn between the way science had investigated psychedelics and the way cultural anthropologists had attempted to understand the Kogi. The first and most influential anthropologist to study the Kogi was Gerardo Reichel-Dolmatoff. According to Reichel-Dolmatoff all aspects of Kogi existence – cosmos, nature, social structure, objects, life cycle, behaviour, aesthetics – are part of the same overlapping configurations ('analogies' is his term for them), creating an overall effect of simultaneous unity.[3] As nuanced as it is, Reichel-Dolmatoff would have been the first to admit that his reading of Kogi culture involved bringing to bear his own ways of seeing, his own 'analogies', that is, in order to 'give structure' (Dolmatoff was a so-called structural anthropologist) to 'exotic' expressions of belief and behaviour so radically different from his own, when the reality had no such structure. It's a kind of reductionism: making difference into something recognisable to us, the raw observable data of Kogi experience caught in a series of crude, foreign conceptual nets. As with psychedelic scientists, the anthropologist assumes that a disengaged, objectifying view is possible and meaningful.

Later generations of 'cross-cultural anthropologists' have begun to consider their own 'filters', to take seriously the extent to which Western ideas about reality might be unable to accommodate, let alone comprehend, the sheer difference of the cultures they investigated: cultures in which forests are 'thinking', rivers and trees are 'like people', in which 'intelligent' plants 'use humans for their ends' – and for whom the very language of 'thinking', 'like people', 'intelligent' and 'use humans for their ends' is never more than a shoddy impostor, mangling reality rather than capturing it, signposting little more than the user's (Westernising) assumptions.

Despite increasingly insistent calls for equivalent nuance and self-awareness, it remains to be seen whether Western psychedelic science can make a similar turn. The implied meaning behind Erik Davis's maxim that 'All psychedelic roads lead south' is that science needs sooner or later to engage with indigeneity, not merely in the form of financial reparation, or promises to 'learn' from 'authentic' practices and rituals, but in the willingness to accommodate a whole world view. To even begin such accommodation requires a sophistication in cross-cultural, interdisciplinary thinking that far exceeds most of contemporary science's set. Yet the briefest interaction with the Kogi clearly demonstrates that any individual 'practice' could only be understood as being continuous with a whole philosophy or culture that was itself inherently psychedelic, a world that demanded constant, dynamic consultation and participation – in the form of payments and divination – rather than static, disengaged observation. One couldn't extract any single thing and hope to make sense of it without taking on everything else.

Daniel read each of us in turn, dropping the *tumas* into the water, interpreting the bubbles, cross-referencing that with a picture of the natural world at that moment, each detail the fibre of a thread weaving our futures.

I thought how this time travel was the indigenous counterpart to Palmer's fevered, LSD-fuelled efforts to bring back something of ultimate reality; and to Sphinx's spectacular drawing down of ancestral energy, like lightning from the skies, for all our benefit. There was no hint of performance here, though. Rather Daniel appeared to inhabit himself more and more deeply as the divination progressed. The initial layer of disinterest in foreigners had disappeared. He was charged with warmth, his eyes affectionate, concerned, meeting ours. Then past us to everything else, as if 'looking at' was really a much deeper form of participating.

It would be Santi who would go first. He had received hundreds of divinations during the years he had collaborated with the Kogi. In fact, Santi's own early experiences with the Kogi were a parable of the difficulties inherent in any cross-cultural 'negotiation' with indigeneity.

Santi was an artist from Cali, his family part of Colombia's landowning class. His father had been killed by a narco-terrorist when he was six. Now in his mid-forties, he had spent the last ten years coming back to the Sierra, working with several Kogi on longrunning projects. He'd made a beautiful film recently about the interrelated practices of weaving, homebuilding and dancing, each one an expression and extension of the unfolding nature of reality. Initially getting access was difficult – there were many governmental restrictions (we'd required a permit to walk into the lowlands of Tayrona National Park today), but the scepticism and reticence of the Kogi themselves were an altogether different barrier to entry. Before they would consider allowing him access, Santi explained, he first had to be taught a fundamental lesson in Kogi philosophy.

At that time Santi was working with large, chemically prepared canvases, which he buried directly into the earth at sites across South America. Each location was considered sacred by different indigenous cultures. He sounded churlish as he told me of his 'artistic vision'. 'My grand idea was that when I retrieved the canvases several months later they would bear the marks and stains of sacred earth; signatures of history, geography, geology – and all that artist crap. "Natural Rorschaches" is how I described them.'

Retrieving his buried pictures from the ground, the issue then became how to interpret what he called 'various shades of brown smudge'. What was it that the Mother, 'Aluna', was communicating? What was being said by her to *us*, for us? Santi certainly couldn't tell. It would, he assumed, require local indigenous wisdom to do the interpreting.

He sought out different tribes who were conventionally considered as being 'close' to nature, to help him read his canvases. Each tribe told him a different story: about climate change, deforestation, financial losses, the threat of war, the general fate of the world – theirs and ours – and what we needed to do in order to avert catastrophe.

Then Santi met the Kogi, who of all the tribes had the most elaborate cosmology. Santi showed them some of his canvases, ones that had been buried in land by the Atlantic coast, at the foot of Kogi territory, explaining his 'vision' to one of the older momas.

' "What the fuck are you doing?" said the moma.' (Santi was not a faithful translator; everything went by way of his foul mouth.)

' "The marks might come from the Mother, but really they are made by you, you fucking idiot. You are just looking at yourself in these pictures . . . It's your stain, just like the one you make with your ass . . . Until you have cleaned yourself properly, all you will see is your own filthy ass." They threw me out of the territory.'

'Years later I went back, and they showed me how and where to do it. I should start with making payments to clean myself. There were lots of stages after that, guided by a moma like Daniel who would make divinations every step of the way.'

It was an object lesson in creativity, in self-awareness, in gratitude, in spirituality for want of a better word. First one must recognise how the world was stained with one's personal view; next one had to learn how to clear that view, honouring one's indebtedness or interdepend-ence on everything beyond the self. Only then could one hope to see the world as it really was. Reciprocity – call it divination or payments – was the means and the compass, the heart of every action.

Daniel looked into the bubbles. He watched the world as it gath-ered itself around Santi. Then he told him that for years Santi had been followed by someone who carried a knife in their shirt. It was time for him to turn and face the lethal stalker.

Next up was Palmer. Daniel dropped the stones into the water as before. Clouds scudded, the wind picked up, blowing sand across the ants' path. Daniel noticed it all.

He told Palmer he must stop chasing 'rare animals'; be satisfied with the ants. Palmer and I shot each other a look: he had been *seen*.

I was next. The first part of my divination was startlingly on point: I tended to be hot-headed and impulsive, Daniel told me. I should sit still and count to four before I reacted to anything. The second part didn't seem as prescient: I should receive the help-stick held out by those who love me. Only later did this take on a potency; either because I finally caught up with it, or because I grew into it, fitted it to my specifications. Daniel looked once more into the bub-bling water before turning to me. 'You will have another child.' Santi translated.

I said 'Thank you', but thought 'Fuck'. I already had two, both teenagers. I had just turned fifty; my back hurt when I picked up a shoe.

Finally, it was Dmitri's turn. We had only met him hours before, a Lithuanian in his late forties who was in Colombia on 'company business'. Dmitiri seemed even more bizarrely out of place here than the rest of us, with his lime-green silk tracksuit, his waxed hair, his spotless white tennis shoes. As Daniel turned to him there was a terrible shriek – the sound of murder – from a bird that had just landed in the palm tree above us. If it had been stage-managed the audience would have found the symbolism overwrought. Daniel's face blackened. He placed his hand on his chest and felt his heart. He spat out aggressively three times, then turned and spoke to Santi. 'This is Shikaka, the messenger or moma bird, a powerful divinator,' Santi explained. 'Her cry means real darkness is here.' Daniel took Dmitri to one side and, with Santi translating, they conducted the rest of his divination in private.

Afterwards Daniel swilled his mouth with alcohol and spat it in the four directions. He gave us each a cotton seed and told us to find somewhere to plant it within the next week – wherever we were, it didn't matter: that way he could look to the earth of the Sierra after we had gone and know something of how each of us was faring. We stood up, shook the dust from our clothes and turned round and round in circles as Daniel chanted. The divination was complete.

Anthropologists have understood divination as a ritualised tradition that involves a dialogue with 'more-than-human' agents who possess information beyond the reach of our normal capacities.[4] In that sense, there is an overlap between ritual divination and psychedelics, with some Amerindian cultures and many New Age accounts of entity encounters all offering potential access to 'special', otherwise unknowable information.

I am not well versed in Kogi culture, but I experienced Daniel's divination in different terms: that, given the right attunement (to use Oberon's musical term) or 'body-set' (to use Merleau-Ponty's), which was at least partly facilitated by the coca (not considered psychedelic but clearly central to both the ritual of divination and to Daniel's felt capacity to see clearly), then everything in the world –

unseen but also seen, supernatural but also explicitly natural – was psychedelic, in the sense that it was mind-manifesting, and therefore capable of being mined for information. You could even push this a few degrees further and understand Kogi divination as testament to the possibility that the world was always reading us.

That still left me with (at least) the idea of 'an unborn child' to look after. I knew something about the Kogi meaning of procreation, not from reading anthropological studies but having met Sylvie Decaillet, an indigenously trained midwife who had been our hostess for a few days in Palomino. Like Santi, Sylvie had spent decades living on the fringes of Kogi culture. Like him, she would describe a pattern of being drawn in by the 'incredible energy' of their culture, then spat out after, because it's *too strong, too different. After a while, it's more than I can take.*' From her unusual access, Sylvie was able to shed more light on Kogi cosmology than any anthropological journal I'd read.

She explained to me that Kogi infants are taught how to make payments to the natural world that correspond to particular body parts: valleys are the spaces between the toes, the head the High Sierra; the Kogi word for dawn is the same as their word for vagina. As they approach adulthood momas will divine a partner for every person (often the women are much older than the men), in accordance with the confluence of their respective lineages. (Given the apparent absence of genetic disorders, I assumed there was some understanding of the principle of consanguinity.) The divination of a partner links the union to a particular place in the Sierra. This becomes the site of the wedding. During the marriage ceremony the man is given his *poporo* for the first time and gets to chew the coca leaves which fuel his desire (the needle pounding the lime inside the *poporo* is an obvious prefiguring of conjugal union). After the ceremony the moma then leads the newlyweds to a special place marked out for them in the territory and they consummate the marriage. Sex is a payment of spirit, but also of bodily fluids. The recipient of the payment depends on the sexual position, as Sylvie explains. 'Orientation of the man towards the ground, the missionary position, is a payment to the Mother. The woman facing upwards is paying the stars and the sky. On all fours is a payment to the animals, the dogs and the cattle.'

The man saves some of his sperm and makes a payment to the earth. Other payments will be made in the form of songs or dance. Later, if the woman has conceived, she will return to the same location to have her child.

Sylvie is one of only a few Westerners to witness a Kogi birth. 'They give birth like mammals, quickly. They are strong, their oxytocin is high, they need no meds. Usually Kogi women don't ask for anything, it happens in silence. Afterwards they have these special techniques for massaging their womb and hearts, which helps the new mothers recover very quickly. The placenta is buried in the ground of the birth site, which is also the conception site. If the child is a female she will return to the same place and offer her first menstruation as a payment. Semen will be offered if it is a male. The female child is then bound to give birth to her own children at the same location. When the time comes for either of the married couple to die, they are buried in the same spot.'

Even allowing for any Western romanticising filter in Sylvie's descriptions (personally I can't imagine being silent during a birth, and I was only watching), it was clear from the basic facts that what we might call nature-relatedness, intentionality, set and setting all informed these threshold moments in the Kogi life cycle – only in a more profound and enfolding way than was ever conceptualised in a Western hospital or academy. And naturally I include myself as being limited in this way. I don't know about having another child, but Sylvie's accounts made me think of the threshold moments in my own life: how dissociated, thoughtless and ungrounded they had been in comparison.

That night we preferred to sleep in the mountains rather than return to the hotel in Palomino. We tied hammocks to the wooden frame of an open-sided storehouse on a ridge above the village. From our beds we could see the High Sierra to one side, the ocean to the other. The night sky was thick with stars and their patterns, the silhouettes of trees, the chirrup of insects – they were all alive in a new way, nature's Rorschaches, laden with possible meaning. Except, like Santi's canvases, we were unable to read anything beyond the stains left by our thinking.

Dmitri opened up a Louis Vuitton manbag, which was really an apothecary's valise, and offered each of us some plant-based delicacies he'd picked up in an 'organic' store in Cartagena. There was *mambe*, a fine green powder made from mixing ground coca leaves with palm ash, several different types of *rappe* which I'd encountered with Sphinx, and some powerful Amazonian plant-based eyedrops known as *sananga* which were meant to help clarify vision. (So painful was the ointment that Palmer instantly renamed it Saramago, after the Portuguese Nobel laureate José Saramago, author of the classic dystopian novel *Blindness*.) The *rappe* was harsh as it hit the back of the head, but if one was silent in the moments after, one could feel it dripping down into the body, opening the heart, cleaning the blood. The *mambe* was subtler still: where cocaine bludgeoned you with the pressure of self-importance, *mambe* was gentle, allowing one to inspect one's own thoughts with clarity, while making the thoughts and feelings of others fascinating and important. (Though I doubted if my friends in NA would see it quite that way.)

The shared divination had made an unlikely group of us. As we lay down, our hammocks only a few inches from each other, hung at different heights for privacy like notes on a scale, Dmitri asked me what had brought me here.

'Ha! A book!' he scoffed. 'Nobody reads fucking books, Dr Andrei. It should be a movie, or even better, a YouTube channel!' His thick Eastern European accent had notes of cockney.

'I can be executive producer – I know all about psychedelics.' His face was stern, unironic, un-surprisable. 'I have first-hand evidence that the plants have saved the lives of dozens of people. And I don't mean "saved" in some bullshit New Age way either. I mean *saved*, literally. I also know they have changed the minds of hundreds of others.' It was possible Dmitri was a research scientist himself, in a newly emerging Eastern European lab, where silk lime-green track-suits were the uniform. Possible, but unlikely. 'And—', with the timing of a natural, 'I also know that I lost my fucking mind to them, and I'm still waiting for it to come back . . .'

With his intro complete he was off, the story of his life in plants powered by *mambe* and *rappe*. At this point in my journey I had arrived at my own version of saturation point, where everything took on a psychedelic hue. But what followed didn't need my help.

It started in Vilnius, Lithuania, in the 1980s under Soviet control. Dmitri was a teenage death metal fan. His parents divorced, his father left home and his mother was sent to the mental hospital with 'heartbreak', a catch-all diagnosis in the Eastern Bloc, leaving Dmitri alone in the apartment, thirteen years old, parentless, no money to pay the bills. He was forced to buy and sell anything he could get his hands on, not much in an impoverished Communist city, to keep his head above water.

Things changed with Gorbachev's rise and perestroika. Suddenly there were serious deals to be made across the Soviet Bloc. Dmitri had quick wits and a fast mouth. Before long he and his heavy metal crew were running truckloads of American cigarettes across the entire territory, from Vladivostok to Minsk. Gradually the gang became more organised, using the code of Zakonniki, literally 'thieves in law', from the Russian prison system. 'One moment we had nothing. The next we were totally stacked. I was twenty years old, I owned a casino, a gym, a house on the Black Sea, and part-owned several underground vodka factories. But it was dangerous, really fucking dangerous. My town went from three murders in a decade to more than a thousand per year . . .'

Dmitri told us about some near misses, like the time he was made to stand in a bath as the police filled it with concentrated acid. 'My best crocodile shoes started smoking.' Only giving up the location of another gang leader saved his life.

In the mid-1990s increasingly 'adverse circumstances' (he'd double-crossed the chief of police) forced Dmitri and his colleagues to flee Lithuania, and relocate to Elephant and Castle, London's own architectural homage to Eastern Europe. Within three months they had taken over a sizeable portion of the black market in cigarette imports in the UK. 'Cigarettes might not be cool and glamorous, but I can tell you they were a hundred times more profitable than drugs or women, a thousand times.' As the de facto leader it was Dmitri's responsibility to look after those of his crew who had to go to prison 'defending our territory'. On one occasion he spent a year on remand himself. 'You can say I went overboard on debt collection.' (They really do talk like this, I thought: the strange loop between drama and reality.)

The trial was cancelled when two of the witnesses didn't appear. Over the next few years Dmitri diversified his portfolio, buying up a gold mine in Angola, and several more vodka factories in Finland.

I repeat, none of us were on psychedelics. The on-my-unborn-child's-life truth was that we were in a remote Kogi village in the High Sierra, barely accessible to Westerners, listening to the psychiatric history of a capo in the Lithuanian Mafia told as a bedtime story, hours after having our fortunes told by a tribal priest.

'But despite all the money and the cars, the women, the houses in Maldives and Dubai, I was totally fucking miserable. Every day I woke up wanting to shoot myself.' The resting state of Dmitri's face was so naturally dour, so over-freighted, sculpted by a lifetime of fight or flight, it was hard to imagine it taking other forms. Sensing his misery, a friend from Belarus told Dmitri about a medicine he had taken in Peru, 'some crazy fucking jungle brew' that had helped turn the friend's life around.

'I thought, what the fuck, I'm gonna try it.'

He took a plane the same week bound for an expensive retreat centre in Peru.

'It was full of people from fucking Malibu, totally crazy, all of them.'

Over the next fortnight he took part in several different plant-based ceremonies. 'They fucked up everything. I saw all the terrible things I'd done. I met my mother in her mental hospital. I hung out with my beautiful son for the first time in years . . . It fucked my head in. Completely. I stopped drinking booze. I couldn't take coke ever again. I even went off shagging for a month.'

Daniel appeared out of the night at the foot of our hammocks, silencing Dmitri. He told us he was on his way to the High Sierra to deliver something to the elder momas who lived above 4,000 metres. It was a seven-hour walk, one way, which meant he would be back here by mid-morning. The Kogi didn't make much of the distinction between night and day, often preferring night for travelling greater distances. Looking out into the mountains above us I had the thought there might be hundreds moving between the sacred sites, making their payments, for the sake of Little Brother, while we lay in our hammocks, snorting tobacco, talking smack. It started to spot with rain. Daniel added some coca leaves to the quid in his mouth and went on his way, walking to the rhythm beat out by the noise of his wellington boots, the moonlight his only torch.

Once again my journey was being hijacked by someone else's, albeit one as gripping and disconcerting as Dmitri's. While listening to him I was also determined to provide some kind of mental sanctuary for the afternoon's divination ceremony. Lying out there in the midst of the vast expanse of the Sierra, the idea of set and setting, of mystical experience, of nature-relatedness, 'popped' in a new way. Everything, as far as the eye could see, in every direction, was implicated in the sense I had of being alive right now, of being sentient or conscious. While this sense rarely even scratched the surface of my heart and mind, it apparently saturated the Kogi world view. That's why even at this moment there would be hundreds of them walking through the night, across the mountains, tending to their territory, making their payments, expressing their gratitude – which really meant honouring the literal ground of their being: the Mother. What for Westerners might be glimpsed through large doses of psychedelic drugs was for the Kogi something like the quotidian fabric of existence. This was another kind of mystical experience, or rather it was a world view consisting only of mystical experiences, and for a brief moment lying in my hammock I had a dim sense of it. Or felt like I did.

I couldn't go as far as other psychedelic authors like Daniel Pinchback and Jeremy Narby, whose odysseys had apparently led them into different versions of supernaturalism or mysticism, but there had been a definite softening (or perhaps reawakening) in me: away from the rational and materialist and towards something less rational, or more poetic, for want of a better word. It made sense to me that the more Western neuroscience speculated on consciousness itself, that 'last big frontier of the unknown' – trying to understand how it arises, what it is made of – the more often it too found itself falling off a similar philosophical cliff. As countless papers attested, purely materialist theories couldn't seem to find a concrete and reliable foothold from which to answer those questions. In many instances, scientific remixes of pre-scientific ways of thinking crept back in to fill the void: like panpsychism, for example, the idea that consciousness is actually in everything, bringing us closer to the Kogi world view, a world view we had spent the last 1,500 years apparently moving on from.

Jaron Lanier had described consciousness as 'a pearl dangling on a thread in all that is'. Only the inevitability of 'observer effects' in any theory of consciousness meant that there would always be 'cracks' in the pearl, meaning contradictions in the materialist account. 'We have to make our minds up about what to fill the cracks with. For some it's more science, for others it's religion. But I chose "kindness". It just seems like the most *helpful* thing given our circumstances.' Even Anil Seth, who has done more than anyone to move the scientific-materialist account of consciousness forwards, and to great success, admits that his convictions involve overcoming strong non-materialist 'instincts'. We find the feeling of being alive hard, if not impossible, to reduce.

In many ways the Kogi ceremony that afternoon was the most trippy, least high high-point of my ten trips. It cracked me open more than any single psychedelic experience. Watching the divination made me understand at an experiential level what the neuroscientist Iain McGilchrist had proposed as an alternative world view: that we in the West were in the thrall of a 'left- hemisphere' view of the world, or 'set'; the spectrum of our seeing had been narrowed, our thinking no longer embodied in the practices and traditions of a community, but fragmented and objectified in ways existentially linked to problems as diverse as rising mental illness and ecological disaster.[5] A different way of seeing was called for (the right-hemisphere view), in which creativity, receptivity to the natural world, attunement to our interrelatedness regained – with or without psychedelic assistance – their rightful central role.

The four of us watched the small figure of Daniel climb over the crest of a nearby hill and then disappear into the night. Dmitri waited respectfully for a few moments and then resumed his story. The man that returned to London was different from the one who had left it a few weeks before. 'I gathered together all my associates and I did something that nobody ever does in the Russian prison system. I told them I wanted out. They had a meeting without me; it could have gone either way, but they decided I could go. I paid some money, they gave me some assurances, and that was that. I walked away a free man. That same day I saw an advert in the paper for a nice-looking private school in York – I thought that was cute, after New York. I put my wife and my kid in the car and off we went. By the end of the week, I'd paid for three years of my son's education, and we were living in a five-bedroom house with its own sauna in the middle of fucking *nowheresville*, Yorkshire.'

Dmitri, paused a moment to fire more *rappe* up his nose.

No wonder, I thought, that the whole of the natural world went berserk when it was time for his divination. Had Daniel really been able to understand the same story we were now hearing, in some analogous form, written into the pattern of the ants, birdsong, the threshing of the trees? It didn't seem likely. But then the bird's murderous shriek, the black look on the moma's face, the removal to a private place to receive the moma's message . . . 'You see now, Dr Andrei, how those psychedelic ceremonies saved my life, and by my quitting my old profession, the lives of many others? But they did more than that. Before I went to Peru, the only meaning I had was making as much money as I could, whatever the cost. Well, that was fucked too. I couldn't do it any more. I'm not saying I turned into a fucking hippy – I still wanted to make money. Once a wolf always a wolf. But everything else changed.'

Dmitri went *green*, green as a silk green tracksuit. With a legitimate business partner, he set up a large recycling operation on the outskirts of Middlesbrough. The new business was grounded in new values: the recycling had to be real, not just cosmetic, its staff – there were 150 within a year – well treated. 'Everyone got free health insurance, extra holidays. We ate our meals together. I never fired anyone.'

Dmitri created his own version of continuing professional development, bringing in an ancestral shaman, Tibetan yogis and New Age life coaches to lead seminars and workshops. This wasn't Mountain View, California, but the industrial heartland of north-east England: they'd never heard of tantra or past lives. Managerial candidates had to meet extra requirements; to join Dmitri for a parachute jump and an ayahuasca retreat as part of the selection process (it wasn't clear if these happened at the same time). He explained how he maintained some of the 'family values' from his previous career, only in a benign, pacific form, spiritualised by the sense of interconnectedness and gratitude he carried from his own Peruvian plant-based journeys.

Such was his natural resourcefulness and flair, such was the good feeling created by his new ethos, that his new business was ridiculously successful, almost as successful as the one he'd managed before. A few days before we met, he'd done a deal to set up a recycling factory in Bogotá, and that was just on a twenty-four-hour layover. The day after tomorrow he would fly to Dubai, where he was setting up three new recycling projects.

'I was a different man. I am a different man. But this isn't the movies – life isn't fucking like that, is it?' Dmitri's dour expression was the stamp of his philosophy. 'Two things went and fucked it all up again, and meant that I've spent most of the last year searching, going all over the world for shamanic treatments, chatting with little holy men in strange outfits on mountain tops.'

The first thing that had gone wrong was an unsupervised experience with DMT in Amsterdam. It terrified him: for months afterwards Dmitri's body stopped being solid, kept spontaneously dissolving, turning into atoms in front of his eyes. He had thought it would help; instead it had left him traumatised. I told him that more and more people were reporting difficult experiences like his. But he wanted help, not consolation. I didn't know what to say, so I told him about my healing experience with Kate and MDMA in Oregon, and he added her details to the file of gurus on his phone. I was now officially part of the hype.

'And what was the second thing?' Santi asked.

Dmitri leaned over his hammock and looked at each of us in turn, as though daring us in some way. The whole Sierra fell quiet, enough to hear the ash from his cigar thud softly on the ground. 'My dog died.'

A beat.

Somehow I knew he was going to say that. Or something like it. Too many third-rate movies? Or a genuine divination? In the spirit of that night I allowed at least for the possibility of the latter, dimly sensing I had my own appointment with death to keep as we headed further south.

9

The Evil Wind

Palmer and I left Santi and Dmitri in Palomino and made our way to the small town of Sibundoy in the Putumayo department of southern Colombia. Set in a valley of steep green hills farmed by the locals, it could have been Wales or Vermont, except the hedge-rows had divided the farmers' fields not into the broad, orderly rectangles we were used to, but into a wonky, inebriated-looking mosaic, reflecting priorities about which I could only speculate.

The town itself was a strip of shops, where car mechanics and bar-becued guinea-pig stalls were over-represented. Nearby, the Putumayo River, one of the tributaries of the Amazon, snaked its way south-easterly towards Peru and Brazil. Santi had sent us here to visit Taita Florentino Agreda (Taita being the Colombian term for shaman or maestro), who served *yage*, the Colombian variant of ayahuasca.

'His medicine is very dark . . . And *muy, muy forte*.'

Santi told us that Flore had healed thousands of people, in the tradition of his father, and his father's father, the lineage going back for centuries. He was the leader slash doctor slash minister of the indigenous Khamzat people, those same people who farmed the Sibundoy Valley with their wonky fields.

Was he a proper Shaman? Was he safe? Was he reliable? We wanted to know.

Santi laughed a little too hard. 'As safe as a stray dog. As proper as a porcupine, a porcupine dressed as a parrot.'

The Psychedelic Renaissance in the West was reflected in the ris-ing interest in shamanic tourism in Central and South America. For many, the venture capitalists and research scientists among them, the pilgrimage south was a mark of their credentials, a badge of

authenticity. Bia Labate, an anthropologist specialising in the pres-
ervation of indigenous practices, told me she receives several calls a
day from 'professionals' wanting a steer on a bona fide shaman. In
the context of an indigenous culture likely to be obscure to these
tourist-consumers, and enhanced suggestibility under the effects
medicine, the idea of grail-like authenticity becomes, she suggests,
a Western fantasy. Meanwhile, tourism apart, the shamanic tradition has 'inauthenticity' – if not downright 'chicanery' – built into it
over centuries of practice. In an early anthropological document
from 1855, the Catholic priest Manuel Maria Alvis notes that Sibundoy shamans 'pretend they have in the woods a tiger which tells
them everything'.[1]

The recent expansion of the psychedelic market meant it hadn't
taken long for Authenticity to become a brand in its own right, a set
of price points for shaman-shoppers. At the more expensive end,
places like Rythmia Life Advancement Center in Costa Rica, beloved
of celebrities and influencers, could guarantee plant-based 'mir-
acles' to 97.23 per cent of retreatants (alongside yoga, bespoke diets
and colonic irrigations etc.), at a rate of $5,000 a week. At the base-
ment end, in Iquitos there were bars where a man in a colourful hat
served the local brew on tap in shot glasses.

But authenticity, like any good, had hidden costs. These are the
words of one Western anthropologist, David Dupuis, who had
witnessed the local effects of the renaissance on consumers and
providers.

I have observed during ethnographic investigations in the Peruvian
Amazon over the past ten years, the regular use of ayahuasca in no
way prevents some indigenous shaman-entrepreneurs from exploiting the natural territories that they occupy in order to benefit their
economic activities . . . [And] while the 'shamanic tourists' claim to
have developed a different relationship to nature thanks to participating in psychedelic rites, my observations tell a different story. In
the long run, their participation only has a very weak impact on
their consumption habits or the modes of production in which they

are engaged, which sometimes directly, and always indirectly, contribute to the destruction of natural resources.[2]

There were also frequent reports of psychedelic retreat centres
exploiting local workers, or the other way round. I heard one story
about a psychedelic spa in northern Peru where the mariachi band
who led the celebrations after the last night of ceremonies had semi-
automatics concealed within their guitar cases. The US guests were
forced to hand over their cash, jewellery and passports at gunpoint
before the retreat's final sharing circle. Even that epicentre of the
miraculous, Rythmia Life Advancement Center, had problems: there
were stories of freaked-out guests being unable to scale the high
barbed-wire fences, while former employees complained of abuse and
misogyny. So it goes. Underpinning our renaissance was a fundamental Western self-deception, a basic psychedelic *naïvety* about the
political and cultural context, enabled in part by the discourse of medical research: that these purveyors of ego-death, mystical experiences,
nature-relatedness and social connection should be immune to the
corruption, malpractice and homicidal tendencies that beset every
other capital trade in the Amazon, and elsewhere; that their miraculous effects should be some kind of guarantee against it.

There was a material backdrop to these depredations: the failure
of Western science and psychedelic corporations to make anything
like adequate reparation to indigenous people for the intellectual
capital on which their institutions were built. Approximately 40 per
cent of Western pharmaceutical drugs are derived from plants that
have been used by indigenous peoples for centuries, and this is by
no means restricted to psychedelics, but includes medical staples
like aspirin, the chemotherapy drug Taxol and many others. The
search for plant-based 'cures' is known as 'bioprospecting'. But in
many cases, this was a charade for 'biopiracy', where foreign agents
extract local resources and knowledge while directing little or no
benefits to the host nations. (Meanwhile, the indigenous people
whose lives depend on the plants lack the financial capacity, or motivation, to fund 'Western-style' analysis of their plants for medicinal

value). Much of this is happening in regions around the Amazon – only a few hundred miles south of Sibundoy – where communities have faced centuries of devastation and oppression.

In order to protect against the threats of biopiracy emerging out of (but not restricted to) Western interest in psychedelic medicine, Bia Labate's Chacruna Institute – named after the plant but also meaning 'Bridge People' in Quechua – has helped set up the Indigenous Reciprocity Initiative. The IRI is a community-led 'biocultural conservation programme'. It works with different indigenous groups across projects including food and water security, agro-forestry and environmental health, fighting for land rights and building economic and educational support. 'Reciprocity' – just as with the Kogi concept of payments – because a disposition of gratitude to natural resources and local wisdom should be at the heart of Western psychedelic medicine. The projects are all self-determined by local people, to meet their own needs and priorities, rather than something imposed from the outside. In other words, in contrast to many Western charities' procedures, there are no strings attached, no specific deadlines or target outputs that are hopelessly tied to the occidental world view.

Taita Flore was 'authentic' by most standards: his lineage went back at least 200 years. He did not have a retreat centre; the medicine was served in his family home. For most of the year his customers were local Khamzat tribespeople who each paid a few dollars for his nightly ceremonies. In the European spring he flew across the Atlantic, where his 'authenticity' was worth a thousand dollars a weekend to Swiss and Israeli seekers.

Determined to distinguish myself from other more callow tourists – less deceived that I might deceive myself better, you could say – I had read some anthropological histories of shamanism. But it was only after Santi had recommended Taita Flore and we had set off for Sibundoy that I happened upon the radical anthropologist Michael Taussig's reassuringly titled *Shamanism, Colonialism and the Wild Man: A Study in Terror and Healing*[3] (I admit it was becoming harder not to think of coincidences as carrying personal meaning). Written in 1986, long before mass ayahuasca tourism, it described shamanic practices and their history in the exact same area I was now heading to: the Sibundoy Valley, 'that valley which for centuries has probably served as a passageway for trade goods and magical power between the people of the Amazonian lowlands and those of the Andean valleys and plateaus of south-west Colombia'.[4]

The sense of good fortune soon turned to discomfort upon learning of the area's colonial history. 'Indeed, we may be told in the Sibundoy Valley that an evil wind issues forth not just from the dead, whether condemned or not, but quite specifically from the dead pagans . . . a temporal hell located in a fermenting, rotting, organic underground of time.'[5] Not quite the history I had been looking for. It seemed I was heading directly onto the set of a transhistorical horror film still charged with the atrocities of Spanish conquest.

'"Where does evil wind come from?" echoed an Indian shaman friend of mine who lives and works in the town of Sibundoy. "From the spirits of the *infidels*, the ancient people that lurk in the earth in certain special places around here that Christians cannot or should not enter",' according to Taussig. The delinquently picturesque Sibundoy farmland was in reality a conflagration of the lingering curses of the 'savage' Huitos and the terror of the Christian missionaries who 'saved' them.

This was the real historical 'setting' of *yage*: not some curated living room in the middle of the hospital grounds where test-tube mystical experiences abound, but the vapoury chaos of colonial massacre. The so-called evil wind, created by the decomposing corpses of the unburied dead – there were no church graveyards at this point – and channelled by the natural funnel of the valley, brought harm to all, but particularly 'to those susceptible people with weak blood'. Here, Taussig tells the reader, 'in the mountains overlooking those "hot Huitoto forests", there were no doctors, no armaments of medical science to do the healing. It was only the highland shaman that had the power to relieve'.[6]

Meanwhile my blood *was* weak. I was susceptible. It was barely a month since I'd undergone my first experience back in London, and it was becoming increasingly difficult to separate all the trips from all the trips. The air miles, literal and metaphorical, had piled up. I had psychedelic jet lag. Rather than heading straight to Taita's house, I had persuaded Palmer to check in to the better of the two hotels in town, where I'd spent a couple of days recuperating with Taussig and guinea-pig take-outs, waiting for my friend Hakan to arrive. And as this is an honest narrative, I'll confess: hating my first experience with *rappe* that night in the Sierra was not enough to stop me from buying a small tub of it from Dmitri, and I was looking forward to adding it to my current stable of pet habits, chewing gum and fizzy sweets (alcohol and cocaine were only pets in the way bears and sharks are) in the relative luxury of my hotel room. 'Anything that makes the heart beat a little faster', as Lenny my sponsor would say.

If Palmer was my trip sitter, the joke went, then Hakan, a programmer who worked in AI for a tech giant, was my 'experimental control'. Born in the suburbs of Istanbul to orthodox Muslim parents, Hakan was the only son of four children. The family would move every few months along the northern and eastern borders of Turkey following Hakan's father's work. It meant changing schools regularly, never settling into stable friendship groups. The teachers considered him 'watchful', 'angular', 'on the quiet side', but 'good at maths', 'interested in new technology'. They didn't realise he deployed it to hack his studies, using MP3 recordings – of choice bullet points and essential facts (when such devices were brand new in Turkey) – concealed under long curly black hair, to ace his papers while his teachers walked ponderously up and down the aisles of the exam hall.

Another thing the teachers didn't know: Hakan's father worked for the government. His job would base him at different military postings where political criminals – often Kurds – were held on suspicion of terrorist actions. It was his father's job to make them talk. He would often come home late, drunk on cheap whisky, hands bloodied from the 'interview', hungry for another fight. Hakan's watchfulness came from those nights and sat alongside his need to take risks. The threat of violence at home introverted him – one might use the contemporary vernacular and call it 'the dissociation of trauma' – preferring to write poetry to playing football or war (another thing his teachers didn't know about). He read some to me once; closely observed, spare, detailing what he saw in the forests and mountains that he liked to hike in, never allowing feeling to occlude the art of seeing.

We had become friends a couple of years before, near the beginning of the pandemic, in the small town of Kas on Turkey's Mediterranean coast. Hakan couldn't swim, was terrified of water, but, having been allowed to work remotely, had come to Kas not just because it was strikingly beautiful and temperate compared to northeast London but because of that same fault-line of fear and excitement – at the level of physiology, with their increased heart rates, blood redirected to larger muscle groups, they were remarkably similar (both make the heart beat a little faster) – that had informed his childhood and since brought him to Sibundoy. The clear, turquoise Mediterranean, free of large predators, made buoyant by high salinity, was a good place to confront his fear. We would spend time in the sea together in the mornings before his working day started. I had no formal training as a swimming coach, but I encouraged him to slow his panicky stroke down, to reach further, kick less and, most importantly, make his breath a meditation: one long, steady, underwater 'out', punctuated by gentle sips of 'in' when his head turned. Over the course of a few weeks, he could swim a *style* of freestyle in a straightish line for the length of the boardwalk, and we became good friends. Two months later we swam three miles together across open water.

Swimming was easy compared to what preceded it. As an immigrant new to London he'd worked three different jobs to pay his way through school, a master's programme, alongside extra courses in coding, only coming up for air with his first lowly post with the tech giant after five long years in a city that lacked the vitality and community of his native country. His parents were proud of him, but they were also concerned: despite frequent encouragements, and occasional set-ups, Hakan hadn't found a traditional Muslim woman to marry. He was thirty-five years old when I met him, as handsome as anything an algorithm might generate, handsome to a comic degree. Attracting women wasn't the problem; rather, Hakan's fretful, restless childhood had left him blighted by the concern he couldn't form lasting relationships with people. He had an attachment problem, to put it clinically. The expectations of his religious family and the contrary pull of Western metropolitan norms added to the crucible of his confusion – conflicting cultural priors, one might say – leaving Hakan unsure what, if anything, he wanted for himself.

I mention all this because it's germane to Hakan's 'set', the complex personal history that he brought with him to the Sibundoy Valley, that motivated him to come in the first place, and would shape his experiences there. Hakan had only heard of 'smart psychedelics' from US colleagues in Silicon Valley who followed protocols with psilocybin to increase their creativity and productivity, in the context of intense professional competition. Otherwise, he was psychedelically naïve, the most naïve thirty-something metropolitan professional left, perhaps. Apart from a couple of lousy cocaine nights on holiday in Cuba he was also drug- and alcohol-naïve. And, rarer still, growing up in a Muslim country, followed by a few years in an urban work laboratory, he was New Age-naïve too. As close to 'a control' condition for a plant-medicine experiment as possible, so the joke went.

Hakan arrived at our hotel the morning after he had done an eighteen-hour flight from London. The three of us took a taxi to Taita Flore's home that same evening in time for ceremony. We arrived at a modest stone family house, a little grander than the other Khamzat homes nearby, located in pasture a mile or so from the Putumayo River. A tin-roof *maloca* had been built onto the home's back wall, large enough to sleep twenty people. The dwelling's modesty was offset by an extravagant garden. A prize collection of roses and peonies neighboured a small pen of pigs and goats and, bizarrely, a de facto open-air aviary of the gaudiest birds imaginable: cockerels, jays, cockatoos, a peacock who preened like a TikTok teen on seeing Hakan emerge from the taxi. And parrots, lots and lots of parrots . . . I'd read in Taussig that, according to local mythology, it was the parrot that had brought *yage* to the taitas.

Flore arrived at 10 p.m. on a small motorbike dressed in work overalls covered in filth. He was a short, powerful man, somewhere in late middle age, we were told, but with jet-black hair and beautiful dark, unlined skin. He shook each of our hands with electric force, jamming a stocky *mapacho* cigar into each of our mouths. Tabaco, to obscure the stench of the evil wind, according to Taussig. '*Cannibales arriva!*' Flore shouted, and, looking at me, 'Head cannibale.'

He greeted his wife and daughters, kissed his granddaughter as though he were tearing into a steak, crashed on a mattress in the corner of the *maloca*, and fell asleep instantly.

Over the next few hours, while Flore slept, a stream of locals came in from the cold night with its evil winds, like characters in a play, dressed in work clothes, tatty jeans, fake designer tops, as different from the dazzling white embroidered costumes of the retreatants at Sonqo as it was possible to imagine. One man had a severe club foot, another wore a home-made back brace. There was a young woman in a shabby red dress arm in arm with a much older man in a shapeless suit and tie, as though they had dressed for evening mass. The woman would spend her entire evening on her knees whimpering, '*Mi Dios es grande, mi Dios es grande,*' while the suited man lay on his back on the cold stone floor, splayed and unmoving like a homicide outline.

Each new arrival prepared their berth, then made an offering to a garish statue of Our Lady of Sorrows, who stood on an altar that was draped in a jaguar skin. Animal bones, coins and wooden whistles lay at her feet. On a bench by the altar sat a small group of Taita's devotees, who had arrived earlier that night from Bogotá. They were educated, middle-class people, among them a young, studious-looking engineer who worked in a large chocolate factory. The next morning he told us, barely able to conceal his excitement, of his idea for a new business combining his two favourite delicacies: Chochuasca, he called it. There was a professional musician who played the Colombian *tipla*. Unlike the unbridled self-expression at the end of ceremony at Sonqo, this man's reverence for Taita meant he had never once asked to perform during ceremony. Then Pilar introduced herself, placing her broken English at the service of any questions we might have about the night ahead. Pilar was an econometrist, educated at the Sorbonne, who worked on a government-backed project to persuade coca farmers under the authority of narco cartels to remove their crops and substitute them for legal alternatives. Not an easy sell. Pilar told us that there was no doubt in her mind that attending ceremony with Taita Flore for more than a decade had changed her world view: she'd gone from being a Marxist atheist to undergoing adult initiation into the Catholic church, an ironic, wonky play on colonialism; indigenous practices Christianising the Western professional.

Shortly before 1 a.m. Taita stirred from his pit and began the business of self-transformation. As he greeted his guests, asked after their families, answered any anxious questions with reassuring directness, he stripped off his outer layer of filthy overalls, doused his hair and face in scented oils and flower water, then slipped a poncho over his Under Armour tracksuit and donned a giant feathered headdress of brilliant greens and reds and blues, made, I assumed, from the birds that pranced around in his yard. He completed his costume by hanging a large pendant of big-cat fangs around his neck, another of miniature ivory skulls, and threaded each finger with silver animal-head rings. He looked astonishing, every inch a taita, except that on his feet he kept the same mud-caked wellies he'd slept in – a badge of indigeneity in this country, apparently – as if to say, whatever form it took, the work was endless.

As Flore rooted around a cupboard behind the altar filled with large metal canisters, I asked Pilar what lineage tonight's songs would be from. Before she could translate, I heard a roar of laughter.

'This is not university! Forget your lineages!' said Taita, emerging from the cupboard carrying a large metal drum which he placed on the corner of the altar. '*El delicioso experimente!*' he announced, teeth as shiny as a motorbike, pouring the dark liquid from the drum into a silver jug.

Tonight was a special occasion: we would try out a brand new batch, '*yage organico*', made of plants from his own garden. 'The honourable cannibals will be my rats!' The whole *maloca* was laughing, apart from us. An experiment, because he had no idea what the strength or qualities of this particular brew might be. In fact, he had no memory of what was in it at all. 'No charge for my guests. Everybody else pays normal price.' The laughter grew more raucous still.

Oberon's church always combined the ayahuasca vine with the leaves of the chacruna shrub (the combination allows the DMT to become orally active), refining the decoction through the lay expertise of the farmers in Hawaii and the feedback of the palates of southern California. The goals of plant husbandry appeared to include making the taste as palatable as possible, maximising visions and smoothing the physical load on those expensively nourished bodies.

Some new psychedelic ventures were taking the idea of refinement to another level, manufacturing ayahuasca in pill form or creating synthetic analogues like 'pharmahuasca', implicitly following the parameters of Western taste rather than authentic indigenous expertise. The danger of refinement, according to some critics, was that it might edit out crucial aspects of the experience itself. Purging, for example, might be considered an unwanted side-effect in the West. Here, in the Sibundoy Valley, it was central. Taussig describes how it is imbued with notions of expurgating the 'infestation' of Catholicism.

Generally the indigenous practice of purging was associated with the removal of 'psychic garbage' from the body, and ideas that might a few years ago have sounded irrational to Western science have recently forced the ground to shift. With the gut having come to be understood as a 'second brain', tied in with cognitive and affective processing, purging has in turn become regarded as potentially therapeutic. While science might consider itself linear and progressive, the reality may be more cyclical, as though it is we who are catching up with ancient wisdom rather than the other way round.

In recent years, psychedelic science has made a show of humility, professing the need to ground itself in indigenous expertise, acknowledging that Western concepts of psychedelics still have only a very primitive understanding of so-called 'extra-pharmacological factors'.

But there's more to it. Even in the more basic science of botany, emerging ethnographic studies point to our fundamental ignorance. For example, Western scientists recognise one species of ayahuasca vine (*Banisteriopsis caapi*), based on the shape and form of the plant, whereas indigenous knowledge keepers know of several different kinds, taking into account not just outward form but also its effects when ingested.[7] Real scientific openness to indigenous world views, indeed a true adherence to the scientific method, will necessitate a significant degree of actual humility in the face of other forms of expertise, and a recognition that our model for how these plants work on the mind is and can only be, like any model, provisional. Above all, perhaps, it will necessitate a willingness to understand psychedelics as more than an instrument of medicine, and as one important thread in the fabric of a whole culture. Even given that willingness, learning is often far from straightforward in practice: for different reasons, resistance to colonial appropriation among them, indigenous expertise tends to be deliberately secretive.

What this meant in practice was that Taita Flore's new brew might contain the combinant plant chaliponga, which was associated with the darker, fermented delinquency of *yage*, rather than the chacruna that was conventional in many Western ayahuasca ceremonies. But it was equally possible Taita had cooked his *yage* with another combinant – one of the brugmansias, perhaps, or any one of half a dozen other possibilities that grew in his garden, including the infamously terrifying deliriant toé. From the way Flore played with gringo expectations, it was apparent that our not knowing (his as well as ours – the delicious experiment was 'double blind') was part of our preparation, just as much as its opposite (knowing and calibrating all the experimental parameters) was the fulcrum of good practice in Western science. Like the higgledy-piggledy arrangement of the farmers' fields, there was a different idea of order here.

Taita was a master at 'edging' our priors: one moment he would impersonate innocence and delinquent carelessness about his *yage*, the next he gave the mystical impression that everything, literally *everything*, was under his control. 'You need many years to understand all this,' he explained. 'Not one night, not one week, more like ten years. Or *ten thousand hours*, as you say.' (He thought nothing of borrowing the jargon of 'expertise' when it suited his purpose.) 'We are the same doctor, you and I' – he was addressing me directly. 'We do many different experiments with our patients to heal the body and the spirit. But there is one crucial difference. In my case I drink my own medicine. Without it, I know nothing.'

There it was in a nutshell: the fusion of first and third person which harked back to the nearly obscured tradition of self-experimentation in Western science, to Sasha Shulgin, to Albert Hofmann and many others, where the complexities of 'method' were less important than the experience of the researcher himself, blind only in the sense that he was drunk on his own product.

'Every experiment comes from things the ancestors have taught us, the knowledge of all the communities, everyone that came before in this area, the whole history of my father and grandfather, the history of this land. I have Khamzat and Inge lineages, different knowledges from different taitas in each community, incredibly rich and diverse. It's all built into this ceremony. But it's also alive in the way we *live* . . . So whenever someone says they are explaining the medicine, they are also joking, because joking is part of the experiment – maybe the main part . . .'

Once again the room filled with laughter.

'*Yage*, the King of Liars . . . !'

I thought of the obscured tradition of Western psychonauts learning to appreciate the tricksterish quality of the psychedelic experience. I remembered from Taussig how Putumayo shamans resist the heroic mould into which current Western image-making would pour them. 'Instead,' he continues, 'their place is to bide time and exude bawdy vitality and good sharp sense by striking out, in a chaotic zigzag fashion, between laughter and death.'[8] Comedy, chaos, colonialism, death and deception . . . a different model of psychedelics was taking shape in my imagination.

The *maloca* quietened down and I told Taita Flore about how *yage* and other psychedelics were being used by Western medicine to therapeutically treat mental health conditions. This was very different from his experience, he told us, estimating that more than 90 per cent of his patients, who ran into thousands over the years (a sample size which rivals the entire population of a decade-worth of experiments in North America), presented with physical rather than mental sickness. This did not mean there wasn't significant overlap in what we both meant across cultures: for while Taita Flore's cohort included presentations like spinal injury, arthritis and cancer among the 'physical', it also included addiction, alcoholism, 'derangement', along with other presentations that in the West we might term 'psychosomatic' or 'somatoform'. But there was still clearly a gulf between our perspectives. Flore didn't really register or acknowledge the different mental health diagnoses as I attempted to explain them; the gap was too wide to bridge. As to the causes of the conditions he saw, they were various, but many of them had at their root 'bad living', the growing influence of consumerism in southern Colombia, he thought, and – most critically – bad diet.

In fairness to the medical model, Flore was equally nonplussed by my accounts of the religious use of ayahuasca in the West, laughing at the idea of building, as Oberon had, a wholesome high-income faith with its Ten Commitments, around something as unreliable, chaotic and deceptive as *yage*. For Flore and the Khamzat people, reverent though they were, Catholicism was all the religion they could stomach. It had become fused with their indigenous culture, so that here in Sibundoy *yage* ceremonies straddled the religious and the medical, upholding and undoing both, while tipping their feathered headwear to the psychonautical third way with their playful delinquency.

Soon after Flore had told us definitively that his tradition was impossible to explain, he started to explain it. Or rather he told a story, a myth of origin that might serve by way of explanation, his suddenly grave tone interrupted only by whispered invocations and sharp staccato exhalations that he made over the dark, foul-smelling brew. Judging by the startled faces of the Bogotá cognoscenti, Flore had never spoken of this before.

'One day, when I was still a young boy, there was a very big gathering around my father, a celebrated taita in this region. A famous Colombian anthropologist was there, who was known for being an expert on indigenous tribes and practices. I remember he told my father that he and his team had discovered the origin of *yage* by using all their technology for dating and chronology. *Yage* went back to this particular ancient tribe hundreds of years ago in this particular area of the jungle and so on.

'Well, I thought, that's good that the experts have answered this question for themselves, because none of *us* who actually drink the medicine knew where it came from. Of course, there were stories about how the parrots had brought it, and we were all very grateful for its being here, but beyond that nobody knew . . . I was by nature a curious child, very curious, and like you I'm an experi-mentalist [he placed the stress on *mentalist*], so I thought, now is the time – even though I'm only six years old – to investigate the origins of *yage* for myself . . .'

Taita paused a moment to recognise the old man in the tatty suit who knelt before him, touching his feet. He stood him up; there were tears in the old man's eyes. Flore whispered something in his ear in Spanish and kissed the top of his head.

'*Muchas gracias, muchas gracias, Taita,*' said the old man, and he returned to the young woman in the red dress who was warming herself with the dogs and cats by a large open fire.

'Where was I? . . . I was sitting at the right hand of my father. I asked him for an extra big dose of *yage* to help my research. My father looked at me quizzically, like he is sizing me up for new clothes. Then he pours me a very big cup.' Flore's eyes bulge as he says, '*MAS MAS MAS grande.*'

'Less than five minutes after drinking, I collapse on the floor – it's that fast. I have died, or, how do you say, I'm in coma, like a dead-sleep . . . That was the moment I first learned about the two worlds: on this side was my body, but my spirit was on the other side. Here it was night-time, there it was dawn. A helicopter waiting for me, not a modern helicopter but a giant, how do you say? A giant blue flying dragon – no, that's not it . . . a dragonfly!' Taita shushed and whistled as the dragonfly took off with the young boy on his back.

'Like *Avatar*,' says Hakan.

'*Avatar*, yes, like *Avatar*,' Flore laughs. 'I see all these amazing things in the jungle: thousands of plants and flowers and medicines, different kinds of tree, cosmic birds, panthers, monkeys, serpents. We fly over the forest, and I see a hole in the trunk of a giant tree. The dragonfly swoops down and flies straight into the hole.' He pauses dramatically. The rest of the 'patients' have been drawn to the altar by Flore's story. 'At the bottom of the hollow trunk is a giant *maloca*, naturally formed, a house of medicine. Sitting inside, on the stumps of fallen trees, were seven ancient ancestors. They were dressed in hats made from beautifully coloured feathers. Otherwise, they were completely naked . . . I could see that their bodies were coated in thick fur, like ape-men. One of the seats in the circle was empty. The most ancient of the elders invites me to sit on the empty seat. Each of the elders is holding a big *toutumo* [coconut shell] filled with delicious *yage* . . . The elders start to dance, strange animal dances. There is singing and shouting. The first elder gives me his *toutumo* full of medicine. I drink it straight down. Then the second elder does the same thing, and then the next, until I have drunk five full *toutumos* of *yage* . . . I arrive at a state of *maximum, maximum* ecstasy . . .'

Flore says 'maximum' with such force that the dogs start barking. He shouts it once again, causing the chickens outside to screech.

'Meanwhile, on the other side, my body is lying dead still on the floor at my father's feet. Already eight hours have passed. My father is worried, because normally it is four or five hours for a journey. He is trying to rouse me, singing and whistling and banging his drum. He makes herbal remedies, burns sage, throws water over my face – still I don't move.' Flore beats his drum for effect.

'Eventually the noise of the drum is enough to bring me round . . . As soon as I realise where I am, back on this side, I'm very upset. I tell my father that I don't want to come back here. I only had five drinks – I was waiting for two more cups from the two remaining elders . . . I am crying and beating my father's chest for bringing me back.'

'Like *Inception*,' says Hakan, but thankfully his audience was still *Inception*-naïve.

'I'm saying, "But Daddy, I know the source of *yage*! I know the men of *yage*! The ancient ancestors from thousands of years before! They had paint on their faces and beautiful feathers, and they were covered in fur, thick black fur, with mouths like monkeys and giant hairy hands . . ."'

He's laughing, Pilar is laughing, the patients are laughing.

'And I told all this to my father in private, because I didn't want the expert anthropologist to hear . . . I didn't want to ruin his story.'

It was a piece of theatre, folklore, a satire on academic knowledge, part of our training in the idiom of reality that came with *yage* nights. At the same time, as with the indigenous lore of purging, it expressed atavistic ideas that Western science was only catching up with. In the last few years it had become common dinner-party fare to venture that ancient psychedelic rites were associated with the origins of Western philosophy, protoscience and religion. Meanwhile biologists speculate on the evolutionary motivation behind our ancient engagement with psychedelic plants: how it might have shaped the brains of our ancestors, fostering cognitive capacities such as advanced mind reading. It also helped us develop, they imagined, a propensity for shared rhythmical movement and early precursors of the art of storytelling,[9] which in turn greatly enhanced the value and appeal of the psychedelic experience.[10] Though these narratives of origin were not much less speculative than Flore's, they seemed prescient. If nothing else, the taita's evolutionary fairy tale of a trip report and general theatrical bent appeared to have whetted everyone's appetite for the night's ceremony.

What followed over those next few hours under the tin-roof *maloca* with Flore and his twenty patients was somewhere between a pop-up theatre, an Irish pub and an ER room on a Saturday night: the sound of men droning the pain of their lives to a descant of the woman in the red dress's bitter sobs, the noise of cataclysmic purging from every possible angle that each time seemed to spark cock fights, which in turn set the dogs on one another, beginning outside then spilling into the human auditorium. Meanwhile, small children wandered through the chaos looking for lost toys. At one point Taita's daughter appeared, dressed in a spectacular ceremonial poncho and feathered headdress, to sing a strange, heartbreaking song, then vanished like a thought. Somewhere in the thick of night all I could hear was the sound of old men in heated conversation around the open fire, as though I had woken in the middle of a NATO summit. One of them ate a sponge cake, another cracked open a tin of beer – so much for the sacred diets of Sonqo. Then the rain came, drilling against the tin roof, making the din more insane still, internalised by those lying down in God knows what ways. We were a long way from Oberon's sensitive symphony, and that was entirely the point. 'In the streaming nasal mucus,' writes Taussig, 'in the shitting, in the vomiting, in the laughter as in the tears, there lies a sorcery-centred religious mythology as lived experience, quite opposed to the awesome authority of Christianity in its dominant mode as a state religion of submission.'[11]

Californian ayahuasca and Sibundoy *yage* were so different they might as well have been different drugs. The *yage* ceremony is a healing, but it's also historically a political act of resistance, a species of religious iconoclasm, against the bloody order of colonial religion. Psychedelic science with its emphasis on individualised therapy in tightly controlled settings has written out this dimension altogether. It stands in relation to the cultural practice of indigenous psychedelics in much the same way as the Catholic Church had. 'In this infinitely warmer and funnier Putumayo world of *yage* nights' (Taussig again), 'there is no way by which shit and holiness can be separated, just as there is no way of separating the whirling confusion of the prolonged nausea from the bawdy jokes and teasing elbowing for room in the *yage* song's irresistible current.'[12]

The one still point, the sun around which this bedlam spun, its organising orgiastic principle, was Flore himself. Except he was anything but still, anything but organised. Rather he seemed to concentrate in himself a greater, more compressed chaos – Essence du Chaos, if he were a French perfume – a rollicking, rambunctious wondrousness. One moment he would sit by the altar banging his drum as though it still required killing – demonically announcing that its skin was that of a real Siberian shaman and former friend – the next he would turn into a female Consuela and whistle discordant melodies without needing to breathe, or play simple looping arpeggios on a harmonica that lasted for hours or less than a minute. At one point I was convinced he had swallowed the harmonica and it had lodged in his throat. This meant the instrument became the conduit for everything he wanted to say, so that whatever bawdy jokes, insults, resentments, curses, whatever evil wind he intended, were magically transformed into the aspirations of a melody, as though the reservoir underneath everything there ever was to say was a heavenly exhalation of sweet gratitude. Then suddenly, without warning, the music would stop as Flore charged outside, and we would hear the altogether different, guttural music of his purging, sung with the ferocity of a dying animal over the wall by the aviary, to the animals who screamed back in unison with one of their own. A single purge would last for what seemed like hours, until the terrible retching would turn, mid-spasm, into more cataracts of uproarious laughter as he staggered back in to the *maloca*. 'Bird food', and 'delicious *experimente*', he would say, the feathers on his headdress newly ruffled.

In quieter moments Flore would visit each of us in turn, chuntering absolutions, spraying alcohol, burning sage, shaking rattles or just consoling. Always at work, always performing, always on the move. Until, silence . . . The deepest practice of all, according to Oberon: managing the shiftless mind without the winch, the kite string of sound. But not quite silent. Because it turned out Flore had crashed out on one of the mattresses and fallen asleep, snoring through the middle of his own ceremony. A few minutes later he was woken by the sound of his own beating drum, his snores turning seamlessly into yet more laughter.

Laughter. A spasm in the non-dual state, according to the philosopher Tom Sparrow. A different kind of mystical experience for me in Yosemite. The sound of profanity, according to Erik Davis, which Palmer and I experienced at clinical levels on our traffic island in San Juan. The main purpose of recreational psychedelics, according to Joe Rogan. A way of resisting the self-importance of the anthropologist, according to Flore. A way of knitting together the chaos of *yage* nights, according to Taussig. And completely missing from Western accounts of psychedelic therapy.

Except I wasn't laughing.

I was suffering, too aware of the uproarious pandemonium around me, because my journey was not all-consuming. The hideous taste of the new brew, this 'delicious experiment', infused my whole night. From the beginning I was anxious about the utter lack of control I had over the carnival around me, my mindful meditation turned to mindless agitation in the face of it. The impulse to run, increasingly familiar as the trips piled up, to tear off into the night and just keep going, and behind it the brute idea that I might somehow exhaust fear itself, that to have any chance against it I needed to be totally alone. No man is an island, but it was an island of one that I dreamed of.

Only this time I'm unable to get up, my legs don't work, have stopped being legs, as though they've had their own ego-death without telling me. Meanwhile my upper half has had a poisonous ego-boost, meaning it has decided that the evening has become 'officially unbearable'. At which point I remember the words of the Kogi divination, holding out a hand for the stick of those who loved me.

The next moment – at least in journey time – I felt one take mine: Hakan, sitting on the edge of my bed, looking at me with radiant eyes as though from paradise. Certain psychedelics used to be known as telepathines: it's possible he had read my mind, felt my need as I felt the need for him. Except now that my friend was actually in front of me, looking at me with those fond, absurdly handsome eyes, what I felt was not relief or consolation but murderous envy. I took my hand away, told him to take another cup.

Whatever he was feeling was mild. The night was just beginning – had he really travelled 6,000 miles for a 'gentle' night? I reasoned. *Schadenfreude* speaking in the guise of experience and authority, secretly hoping a higher dose would re-route his journey on a darker orbit . . . But underneath all that deceitful, malicious crap, what I really wanted to say was . . .

'*Help!!!!*'

I was not wrong: Hakan, my 'control', was indeed in paradise. My sending him for another dose did not condemn him to some version of the infernal, but catapulted him instead into much deeper, stranger bliss, as I would discover the following morning. While I've experienced marvels under the psychedelic spell, I have never had such uniquely, iconically miraculous experiences – crammed with all the gold-standard signatures – as Hakan, a Muslim computer-programming-psychedelic-virgin, had that night. And it was all my fault. He couldn't stop thanking me! They say you get the trip you need, not the trip you want. Well, apparently, I needed a lethal dose of trip envy, and Hakan needed heaven. Here's what he says about it in the journal he shared with me, written in English, a language he'd only learned in the last few years.

Tired of endlessly thinking about the past, I decided to set my intention on the *future*. What should I do with the rest of my life? Where am I going? Who should accompany me on the way? These are the questions I have in mind. But I'm also determined to be obedient to the ceremony, to accept whatever it is I might be shown.

After purging the first horrible shot within ten minutes, I felt a sense of relief. I was so grateful to be here. I went over to Andy to thank him for inviting me. I felt like I was made of gratitude . . . But Andy looked tortured, I felt as though I could see a lifetime's suffering gather in his face. He told me sharply to go back for another cup, but I trusted his judgement, I was grateful for his wisdom. It turned out to be wonderfully prescient.

Even though I purged again a few minutes later, it was time enough for the medicine to make it into my body and set me off.

The whole jungle lit up in an instant. I feel something formidable and mysterious giving birth under the stillness of the forest . . . It feels like every single creature in *maloca* – all the animals and sick people – is giving birth and being born at the same time; that we are all joined. No separation, no death; only birth is happening. And it feels profoundly safe, in a way things have never felt safe before.

Normally Hakan's speech was matter-of-fact, but I could hear the childhood poetry being drawn out of him from under the prose of his second language.

'My hammock had become my chrysalis. Every shape on the roof, every log on the fire, turns into an animal spirit: wolves, tigers, pumas. Each one symbolised a different feeling, a way of being free, without the expectations of others, that I've never let myself experience before.' As mentioned, much of Hakan's mental life was the crucible for familial and cultural expectations.

Now, for the first time, I know freedom is always there, inside, waiting if I need it. In the different animals I see strength, persistence, agility, awareness, purposefulness, beauty, and cruelty . . . I sense that the bandwidth of my imagination is widening, that all my feelings have been so terribly narrow up until this point.

I look out through the window into the jungle, which is veiled by moonlight. Alongside all the noises and images I hear its voice: the voice of the jungle, the voice behind everything. *Buy a part of me and protect it. Live in me*, says the jungle.

I've heard many people say that 'the plants' have told them something: to leave a partner, to change jobs, to take up painting, to invest in palm oil. (One person in San Diego said that Aya had told them to get a boob job.) I was circumspect about most of it. But then I was also jealous: I had never been told anything so concrete, and the child in me longed for instruction, however absurd. (Just so long as the instruction wasn't as absurd as being told to have another child.)

It felt like the whole purpose of my life was unravelling and coming together at the same time. The words of the jungle made total sense: I could secure myself only by choosing to protect something outside and beyond me. The jungle needs to be cared for. We need each other.

Then a girl appears. It's Meric, a Muslim girl I met in a coffee shop in Kas last year. At the time we got on well, but she left town the next day. We messaged for a few weeks, talked about meeting up again, but it slowly faded. But now she was back, standing before me. Suddenly it is very clear to me that Meric is the *one*. I will look after her, she will bear my child, we will live in our forest. It's all so simple, so compelling, so real. I asked these questions in my intentions for tonight's ceremony. And they have been answered clearly, thrillingly.

The excitement is enough to get me out of my hammock. I walk through Taita's garden into a small field of farm animals. I can still hear the music coming from the *maloca*. I have this deep sense that everything in my life has been totally resolved. I'll resign from the company. I'll buy land somewhere in Turkey and marry a Muslim girl, Meric, or someone like her. My parents will be happy, I will be happy. Everything in life is coming together perfectly.

I look down and suddenly the ground is flying past me a hundred metres below. I realise I have turned into a giant bird, like a condor maybe, though I've never seen one, but it's a huge predatory bird, anyway. I am flying over the jungle; on my left wing I am carrying my family, and on my right wing is Meric. I have found the love of my life; my family is secure. I am a giant condor. It is amazing.

Suddenly the journey jumps forward to the night of our wedding. Meric and I are naked in a jungle glade after the ceremony. I am deeply attracted to her. She is so perfect. My penis grows much, much bigger than it is normally – a foot long. It keeps growing, bigger and bigger. It's pointing straight at Meric, as if to say, *She is the one*. But then it curls upwards and starts to point back towards me. It's still growing, only now its heading for my face. I have this terrible sense that the penis is heading for my open mouth. The journey is turning into a nightmare, my worst fear: I will have to swallow my own giant penis. But just as it reaches my chest, the penis turns inwards and goes straight through my rib cage and plants itself deep in my heart, like a tree . . . Now I am having sex with, no, I am making love to my own heart. It's pure bliss . . .

Time jumps again. I'm flying in the sky, a condor again, the earth miles below. I see Meric and my family holding on to my wings. All of them are vomiting. Each time they vomit they shrink in size, more and more, until they are like little insects holding on to my feathers, until they can't hold on any longer and they fall off. I am free to fly over the jungle, soaring on the music . . . I open my eyes and I am back in the *maloca*. There is the taita, standing over me in his bird outfit playing the mouth organ, drawing me into his heart with the music.

Meanwhile that same night, on the same medicine, exposed to the same pantomime of pandemonium as Hakan, I fell deeper into the abyss of fear and envy. Taussig puts envy centre-stage of the epic theatre of Sibundoy's '*yage* nights': the lethal desires, the failures to reciprocate friendship, treasons both sly and crass, all manifested in 'the discordant symphony of purging'.[13] And purging is also envy's antidote, the forced eviction of those hot, hateful splinters lodged deep in the body, what he calls a 'spuming of the social bond itself'. (Apparently the taita's constant infernal 'shushings' and 'whooshings' are to draw envy out.) Envy: the opposite of gratitude with its necessary humility, dependence and submission; the black mirror of Western psychedelic science's 'interconnectedness'; the dead spot at the heart of the myth of the self and its fantasy of individual freedom; staked to the delusion that we can do without others.

I could not vomit. The cramp in my stomach was so bad it felt as though an amateur magician had accidentally cut me in half. I knew what it was like to feel cancer moving through my bowel, as my father had before he died. Then I knew what it was like to be my dying father. Then I knew what it was like to be the cancer moving though my own bowels. I longed to purge. I longed to die. Then I was longing itself, all of it, everywhere, at the same time.

Erik Davis had told me that sustained psychedelic practice, like Buddhist meditation, was really a preparation for death. Many of the psychonauts he knew had met their own death with remarkable equanimity. Roland Griffiths, the research lead at Johns Hopkins, had recently been given a terminal diagnosis. He continued to meditate and practise with small doses of psychedelics. In a podcast with Sam Harris, remarkable for its intimacy, he described how, rather than fight his diagnosis or regard it in catastrophic terms, it has afforded him an incredible, pellucid frame through which to experience his remaining time, saturating his days with gratitude for the miraculous preciousness of being alive. Death was making the remainder of his life, he tells us, feel intensely psychedelic. I was not up to this lofty perspective. My 'frame' was its negative: I saw the inchoate longing that had shadowed much of my life, as really a longing not for wealth, success, love, but to die. I remembered the first part of Daniel's divination: that, however strong the urge, I must resist it four times. So I waited out my own death alone, without vomiting, without functional legs, while the evil wind blew all around me.

People die all the time on psychedelics. I'd heard it said time and again as I bounced around ceremonies: that the medicine trains us in the 'art of not knowing'. If that's true, then death is un-knowing's *coup de théâtre*. I opened my eyes and Taita was holding out a cup of *yage*, my mortal cocktail. I swallowed the smallest drop of the horrible brew and died instantly. Yes, I died. Often people mean it metaphorically, poetically, dilutely, but my death was medical, cold, hard as a mortuary slab. My body turned white, unmoving, losing the last vestiges of warmth. I felt death moving in the roots of my fingers, under the soles of my feet, on the inside of my bones. There was still the glimmer of consciousness, strong enough to illuminate death's shadow passing over me, death's taste. There was no escape, nothing to be done other than accept it, again and again, until it had its fill. Then I got to my feet, my legs reborn, went out into the night, and vomited to my heart's content over a low wall. Boy, it felt good to be dead.

There was a tenderness in the air the next morning. With the rest of Flore's patients we cleared up the carnage of the previous night, each person silently sifting the debris of their inner journeys. Hakan basked in the 'aya-glow', scrolling through his phone looking for 'jungle' to buy in northern Turkey, phoning his father to tell him he was ready to be head of the family and find a decent Muslim wife. Later he would tell Palmer and me that he was determined to steer his Big Tech bosses to more wholesome practices, help ensure their algorithms were better aligned with less rapacious human values. While Palmer supported the sentiment from the bottom of his heart, bitter personal experience from a similar place led him to question the wisdom of such a gambit.

Research suggests that the psychedelic experience is frequently marked by the striking feeling of having obtained unmediated knowledge, without need of external validation or evidence.[14] Whereas for Taita Flore and the tradition of *yage* nights in Taussig, ayahuasca was the great liar, more in keeping with another tributary of research in cognitive science which pointed to the relative ease by which we are persuaded by false insights and fake eurekas. So what to make of Hakan's visions? Where New Agers would talk of direct messages from Mama Aya, neuroscience understood them as 'stereotypisations' informed by cultural priors and 'acquired knowledge',[15] meaning that they're shaped by the more or less conscious knowledge or beliefs we have about psychedelic hallucinations. Hakan was psychedelically naïve; he had no prior experience, and to my knowledge almost no acquired knowledge (*Avatar* apart). His hallucinations were a splash of Disney-level South American lore and symbolism added to his native Muslim culture and beliefs. And yet his journey had classic elements: the jungle speaking, the flying condor.

The research into the socialisation or contagiousness of certain hallucinations, how they might shape the emotions and the categorisation of perceptions, is only just beginning. Personally, I reliably have botanically themed visions on ayahuasca, but never have them on smoked versions of the same DMT, even when that DMT is extracted from tree bark. As with Hakan's experience, accounting for this remains enigmatic.

Flore reappeared from his kitchen, where he'd been holding a confessional with each of his patients while breakfasting with his family. This was the opposite of a church or retreat centre. Taussig notes that in Sibundoy the daily co-existence of 'patients' and the family in the shaman's house 'demystifies and humanises, so to speak, the authority of the shaman'. With his mouth still full of eggs (eggs made from chickens reared on the vomit of *yage*), Flore told us that he stood in urgent need of our help: last night's rain had caused the Putumayo to breach its banks and flood a Khamzat territory three miles to the west. We longed to be useful, and he knew it. As it turned out, our trickster taita would make a subversive joke out of the pieties of Western gratitude.

We drove at speed in a new Toyota Hilux – the European price of 'authenticity' – through the Khamzat settlements, Flore making several abrupt stops to chat to the shoeless, the elderly, the infirm, handing them pesos from a pile of notes stuffed into a mug on the dashboard. Grateful, they blessed their taita with sign of the cross. We drove past a site where giant municipal diggers were trying to reroute a section of the river by creating a new tributary. Further on we arrived at a bridge which had collapsed under the flood, marooning dozens of families on the other side of the river from their homes. A group of a dozen men in deep conversation broke off, seeing Flore, and swarmed him with their concerns. How different he's become again, I thought, as he listened with gravity in silence, the still point among these shouting men, these grieving families. Then he returned to us and said he needed our help. We took off our jackets, rolled up our sleeves, readying ourselves for whatever digging or carrying he had in mind.

'Three hundred dollars each.' The taita held out an empty, matter-of-fact hand.

There's no such thing as a free ceremony, I thought, as each of us coughed up all the dollars in our pocket. This was Flore's version of the Indigenous Reciprocity Initiative. The grateful villagers gathered around us for a series of group photos.

'You are Chacruna. It literally means Bridge People. The villagers say they are going to name the new bridge after the gringos.' Torrents of tribal laughter.

Back in the *maloca* it was time to close the ceremony. *Rappe* was the local version of the 'trip-killer' – the antidote we would take to draw our journey to its end. Flore, dressed once again in the magnificent parrot headdress, fired the tobacco ash deep into my brain with his three-foot-long blowpipe. One moment I was looking into the jewels on the ringed fingers that gripped the pipe, thinking, 'Wait a—', the next: '*SSSHHHHHHHHTT!*' It actually sounded as if something was being *sucked out* of you, something important. My mind dropped into my feet like a lift whose cable had been cut. It was . . . grounding, and I stepped into the garden to breathe fresh air. A few moments later Hakan crawled past me out of the *maloca*, arrived at the low wall over the chicken coup and vomited over it.

'That was Erdoğan,' said Hakan, looking at his insides below.

It started to rain again. The evil wind had gone. Believe me when I tell you that I felt nothing other than compassion for my friend as he lay in the same spot for the next hour, groaning, writhing, unable to move. We went to find the taita for help, but he'd vanished.

Eventually Flore showed up, hair glistening, bare-chested, fresh from a swim hole at the bottom of a nearby waterfall. He laughed, of course, when we told him what had happened to our friend. Then he prepared an antidote to the antidote, which, to Hakan's horror, was *more* of the exact same *rappe* that had floored him an hour before. The taita blew down the same pipe, the same screamed whisper, the same deep groan, only this time within a matter of seconds Hakan had sprung to his feet and, wiping the spit from his chin, told us he was ready for kebab.

10

A Psychedelic Century

It turns out Hakan, my drug-naïve experimental control, had already been corrupted without my knowing.

I drank a small cup with some friends at a ceremony in this guy's house. It's the first time I've done this. Three hours have gone by, yet I could not feel its effects. This could probably be one of those fake teas.

And now, I'm feeling it! I can feel how my body is getting heavier, it also feels weightless as if I'm in outer space, floating out of my lungs and in the nothingness. My limbs are feeling itchy, and I can't stop twitching.

I see beautiful jungle scenes as though they are real: snakes, jaguars, multi-coloured birds, a giant condor . . . I feel that I can see my soul in the shape of the universe and I am one of the trillions of atoms floating.

Ten hours have passed, but these are no ordinary hours. After being immersed in the wormhole of time, space and matter, it has given me insights about the nature of the universe. But I am a human being with a human brain, and it is impossible to store all of what I have learned that night. For the first time in my life, I wish I do not sleep and continue to learn. Maybe I could be an immortal being, an all-knowing deity.

I close my eyes in awe as I embrace the journey of getting high on sacred medicine. I wish to be on this journey for the rest of my life. But even I need a break.

I woke up. A day has passed. After calming down, I realise that my insights from the trip only confirmed my belief that God doesn't exist. And this has given me a different kind of closure. My first experience has been positive and left me thinking how wonderful life is. It's true that it helps you to see things in a new way.

Not Hakan's words, not Hakan's previous experience. It wasn't even a human experience (never mind its 'human brain'), but an early (mendacious) language AI ('*Even I need a break*'? . . . But do you? Really?). Waiting for a connecting flight at the airport in Bogotá, on our way to Sibundoy, Hakan had asked an early chatbot (this was 2022) to tell him its 'experience' of ayahuasca. The jumbled Wikipedia montage of a trip report was the likely source of his hallucinatory contagion: the beautiful jungle, the various flora, even the giant condor . . . It made me think that the growing need to confer with GPTs – contacting some strange unseen beyond – was really the latest version of schlepping halfway across the world to consult an indigenous shaman. Erik Davis thinks of the current trend as the beginnings of 'cybernetic animism'.[1] Meanwhile the computer scientist Gary Marcus describes the faulty simulacrum of intelligence displayed by such chatbots in terms of '*hallucinating* the space between data points'.[2] The machines themselves are effectively tripping without knowing it.

The three of us – Hakan, Palmer and I – made the short, 200-kilometre journey from Sibundoy to Narino, a village in the department of the same name which stood in the shadow of Galeras, a 4,000-metre volcano listed as Colombia's most active. There, in a wooden A-frame, nestled in steepling fir trees, was the home of Alejo, a young Andean taita. He was a mestizo in his early thirties with a waist-length ponytail and tired, baleful eyes. He wore the local mountain uniform: a green feathered fedora and a white alpaca poncho so fluffy and voluminous it looked as though the animal was still alive and carried the diminutive taita around half-swallowed.

My last trip. Honestly, it felt like one too many, my bucket list filled with plant-based purge. But there were reasons to keep going. For one, I'd never tried a mescaline-based psychedelic. Reading Huxley's trip report in *Doors of Perception*, I'd marvelled how the drug had, even in the most quotidian of settings – the back garden, the magazine section of a store in West Hollywood – brought him thrillingly close to the transcendental qualities of the natural world, painting, music and . . . *garden chairs*. (Less supernatural revelation than the confirmation of Huxley's mystical 'set', mescaline the cosmic icing on his perennial philosophy.) There was also the thought that, for all the talk of the Psychedelic Renaissance, an industry reborn, the indigenous serving of *wachuma* (the South American mescaline-based tea made by boiling down large hunks of San Pedro cactus) was a dying tradition, on the brink of extinction, making the last leg of the journey south feel like a race against time.

More intangibly and more personally, my journey was a series of case studies on the drugs, but also neat little studies of the people I'd met along the way. All the people apart from me, that is. I was there throughout, of course, but other people kept having more compelling, more significant, more transformative experiences than me. Something personal was missing. I had the sense that some kind of reckoning needed to take place, that some 'beast in the jungle', some 'unborn child', a clarifying scene or clinching diagnosis, must still lie in wait on the fabled path south. So, onwards!

I had written to university departments for information about indigenous lineages using *wachuma*. Bia Labate – who set up the Chacruna Institute to promote psychedelic justice for indigenous traditions – had told me, though she couldn't be sure, that, excepting the Q'ero tribe in Ecuador and a few 'delinquent' cultures in northern Peru likely to be no more than a hundred years old, no such indigenous lineages remain. I heard back from one anthropologist in Bogotá, who had 'sat' with Alejo in ceremony a few months before. He had no idea about the provenance of the young taita's lineage, but he could certainly vouch for the strength of his tea.

Though formally uneducated, Alejo told us in quietly spoken, melancholic English, that he had become a de facto anthropologist, spending much of the last decade scouring the continent for what remained of the tradition. With his research, like the cultures themselves, all but finished, he was – at the insistence of many elders – tentatively conveying his knowledge. We were among the first Western visitors to have been to his small, rudimentary home. He introduced us to his wife Aleja, his mother and four young children, who were making mini Galeras out of their dinners of rice mushed with plantain and avocado, and led us to a small area on the landing outside the family bedroom, sheeted off with a cotton drape; our quarters for what he hesitantly called 'the retreat'.

The mood was sombre around the dinner table that first night. The talk was of the upcoming elections. The incumbent president was on the extreme right, his government corrupt, patsies to the narco-cartels. It looked increasingly possible that the candidate from the extreme left would win, which might, Alejo thought, make matters worse. At thirty-two he carried the elegiac air of someone who had seen it all. In the West the talk was of the need to create 'positive expectancies', of 'placebo-enhancers', but this taita seemed utterly unconcerned with impression management. The situation wasn't *positive* – people were beleaguered, traumatised, desperate, he told us. They were also fickle and historically naïve, he added, which left them free to tell themselves whatever story they wanted to hear, changing it the next moment according to their fancy.

Just as we were making ourselves comfortable listening to his self-damning native line, Alejo turned on us. Were we any different, wealth and education apart? No doubt we took a certain voyeuristic pleasure in South American extremism, but was politics really different anywhere these days? The world was all but equally divided into democracy and authoritarianism, and even those differences were narrowing. He spoke with the fluency and sealed intelligence of an autodidact, his tone more stoic than spiteful. And now, Alejo continued, research scientists had the audacity to bring psychedelic medicine into this farrago, citing a piece of research from Imperial College that made the 'crazy utopian' claim that psychedelics would liberalise people. 'Who paid them to do this?' he wondered. 'The government? Google?'

The idea defied common sense, he contended. If liberalism was the outcome of ceremony, then how was it that Colombia and Brazil, who collectively drank more psychedelic tea than the rest of the world combined (Alejo's mother had started him on *yage* when he was a baby, a practice that was not uncommon in this part of the world), had some of the most extreme right-wing regimes on earth? This was the problem with so much of the new research, he told us:

it was 'politically ignorant, culturally stupid', thought up in expensive laboratories by men with 'heads halfway up their asses'. For all the high talk of ego-death and miraculous transformations, 'situations remained the same, people remained the same.' This was a psychedelic retreat in reverse; it began with the come down. Our taita was hype-proof, to the point that I found myself wanting to push back a little, make a case for small signs of improvement, a broadening of sensibilities. But more guests had just arrived. It was time to head out.

We joined a circle of a dozen young men and women from the nearby city of Pasto sitting around an open fire, dressed in fedoras and ponchos, for the evening's *'mambedero'*. This, as I understood it, was a free-flowing teaching given by Alejo and inspired by *mambe*, the mixture of coca leaves and ash we had encountered with Dmitri in the Sierra.

As night fell, the stars scintillating more brightly through the haze of volcanic ash, Alejo told the heartbreaking story of his time travelling around remote South American hinterlands, harvesting what he could from the disappearing lineages – songs, brewing techniques, cosmologies – as though they were the clippings of dying plants. He spoke of the 'majestic' Kallawalla, healers from the mountains of Bolivia, the Huaringas, the 'light-footed' Q'ero from Ecuador, the 'bedevilled' Aymara and Chavín from northern Peru. He told us stories of the Masters, of Ruben Orellana, Agustín Guzmán, Oscar Herrera, Omballec, and mentioned others whose names and lineages he was bound to keep secret. I say 'told', but his words had the rhythm and lyricism of songs: one verse dramatic and exuberant, the next shattered and self-effacing, always gravitating around the same tragic kernel. This was the twilight of *wachuma*; without Alejo and a few others like him its rich, magnificent history would die unsung.

'The earth, Pachamama, offers up different medicines to treat the ills of the Age.' As he explained, tobacco, coca, *yage*, *wachuma* and *wilka* (a snuff containing both DMT and 5-MeO-DMT) each constituted one of the five main branches of the 'Tree of Wisdom', and were medicines for specific moments in history. Ours was the Age of Consumption, afflicting us with a reckless, insatiable hunger for everything that lay outside ourselves. *Wachuma* was the medicine we needed now. Then, making eye contact with each person, Alejo pronounced that we, all of us, would be its torch bearers.

While I might cavil at the occasional New Age condiments that seasoned his stories, and certainly at the methodological rigour of his fieldwork, there was a supple intelligence at work here, which made the narrative that much more urgent and compelling. And strange, too, because in reality I had no way of knowing if anything Alejo told us was true. One thing I had become aware of in me these last few weeks was a general increase in tolerance of the unfamiliar, a relaxing of my instinctive scepticism. It was hard to evaluate this small, but significant shift. Was it a movement towards the natural faith in not-knowing that the psychonauts spoke of? Or was it a trait shift in suggestibility after too much Kool-Aid? That was the trouble with first-person research: there was no control. But one thing I felt sure of: whatever the truth of his story, there was something manifestly authentic about Alejo, a sadness that chimed with the tone of the world he described.

The young men and women in the circle that night were not the agrarian poor and infirm of Taita Flore's cohort. They looked middle-class, healthy, secular. Among the group were two of Alejo's cousins and a young tousle-haired teenager, Alejo's nephew, who was dressed all in black and preferred to keep himself in the shade of the tall trees outside our *mambadero* circle. As Alejo's story wound down I turned to the translator to ask what had brought the rest of the group here tonight.

Alejo's oldest cousin spoke first. He co-managed a business with his younger brother, sitting opposite, that specialised in the rehabilitation of children with neurological damage. This much was in keeping with the retreats I'd been to in North America, where the 'helping' professions formed a significant proportion of the retreatants. But here the job description was just a clearing of the throat. When he was not at work, the older cousin, a handsome, bearish man in his early thirties, led an organised cabal of football hooligans associated with the team in Pasto. It was a semi-military operation, running into the hundreds, part-funded by one of the narco-cartels, and specialising in drug-running, racketeering and robbery. The young man sounded like any besieged CEO: overwhelmed by logistics, unable to exercise control over certain insubordinate factions, kept up at night by demands from above. His 'work' with *wachuma* had only just begun, he told us, but it was already clear that the medicine was making the criminal life deeply unpalatable. It was hard to feel all this connection – to oneself, to nature, to other people – and then stick a gun to someone's head. He sat down damnedly.

Then his younger brother on the opposite side of the circle took to his feet, a smaller man with luscious hair and matinée-idol looks. He too was a gang leader, but his was a national-level organisation, meaning more logistics, more politics, more responsibility. His brother's gang was the sworn enemy of his own. Even though their civilian jobs brought them together every day, they were mortal enemies. This would be his first experience with *wachuma* – he'd come only at the insistence of his girlfriend, and was not looking forward to it. He flashed his brother a filthy look and sat down infernally.

Over the course of the last twenty minutes a membrane of hatred had formed and bisected our fireside circle. One by one each person stood up, declared their allegiance to one side or the other, and spoke of a life mired in gang violence, criminality and remorse. If Taita Flore's group were as far as it was possible to be from the Bitcoiners and professional non-dualists of Sonqo, then Alejo's gang leaders were a third point on a triangle – equally far from the other two.

As the circle broke apart and people retired to their tents, I asked Alejo about his nephew, the boy dressed in black, the only person who had not spoken. Jorge was sixteen, he told me, and had a diagnosis of autism. He rarely spoke, but would always turn up at the house when Alejo was holding ceremony. Jorge lived for the cactus tea, the only thing he had ever encountered that made him feel anything. Alejo told me it gave him a 'wildness' inside, 'like a screaming monkey swinging through the jungle of his head. He'd drink it every day of his life if I let him.'

The next day, Good Friday in the Christian calendar, Alejo planted us in a forested hillside facing the volcano. Twelve people spaced regularly like lamp posts but out of sight of one another, and all within a few hundred metres of the *maloca*. My spot was in the midst of three neat rows of fully grown ancient-looking San Pedros, their trunks forking high above my head, and beyond them woods of yew and eucalyptus.

As the *wachuma*, another lighter brew of acrid tea, seeped into my nervous system the sense of others being close by disappeared. I sat there without moving a muscle for several hours until Alejo and his helpers exhumed me with the sound of a candle-lit drum and the smell of sandalwood incense wafted by a condor-feathered fan. By then, day had slid down the stone slope of the volcano and it was night again. It was as though nothing had happened, but in the most intense way imaginable. It's startling how clear nothing can be: unobstructed, straightforward nothing.

Empty myself, I became the matter-of-fact recorder of everything – literally everything – that was happening around me. Nothing was lost: every creak the wind made in the branches, the shapes made by the light as it streamed between the leaves, the sputtering flight of a bumblebee. But as the day waned, my view of it shifted, as I realised my eyes had been shut the whole time. My fine-grained 'observations' had had nothing to do with recorded reality at all. Instead, they were a series of meditative fantasies on green, *in* green, that imperceptibly loosened themselves from the natural facts of a Colombian hillside – there were no bumblebees here – and replanted me in a large manorial garden in seventeenth-century England in the late afternoon, in that philosophical dusk before reason had fully established its empire, when, in the words of the writer Keith Thomas,

for most people . . . the plant world was [still] alive with symbolic meaning. Certain trees and shrubs – rowan, vervain, mistletoe, angelica – were worn or hung up to give protection against witch-craft. Others – bays, beeches and house leeks – were planted near the house to save it from lightning . . . Crucial to these practices was the ancient assumption that man and nature were locked into one inter-acting world. There were analogies and correspondences between the species, and human fortunes could be sympathetically expressed, influenced and even foretold by plants, birds and animals. Hedge-hogs, swallows, owls, cattle and cats, all gave out signs of a future change in the weather. Sailors watched for the stormy petrel, while the housewife used the cricket on the hearth as her barometer. If the ash was out before the oak, they were bound to have a soak. Worms in oak apples presaged a pestilential year.[3]

Somehow in this Andean cactus garden, under the influence of *wachuma*, I had happened on a lost 'Englishness', an indigeneity of my own – a time when everything in the natural world was still charged with urgent, personal significance – layered in me as deeply as the land, only obscured under the mud of hundreds of years of reason. The world was enchanted, and we had disengaged from it: that was the lesson. Hakan would, I imagined, have an Ottoman version of the same, Palmer his own Dutch or German variant – whatever culture preceded his family's move to the States.

The next moment a distinctly un-English porcupine with a pig-gish snout, a long, spiked tail and a sense of irony scuttled by my bare feet. It felt poetic, poetry being after all nothing more than 'a real toad in an imaginary garden', according to Marianne Moore.[4] And these self-conjugating thoughts – these psychedelic confusions of fantasy and documentary, of identity and history, of nature and mind – were their own poems of sorts. And like a poem, the unlikely porcupine vanished.

This was the gift and spirit of *wachuma*: a deeper, visionless wife to *yage*'s gaudy show, day to night, her *this* to his *that*. Except, as if to say, 'I can do that too', the cactus offered up a towering spectacle of its own. There, nailed to a lone yew, was the body of Jesus Christ. Only there had been an uncanny reversal of the crucifix: the dying tree was clearly nailed to a living man with outstretched arms. Christianity: 2,000 years of getting it the wrong way round. *Oh, Wood of Sorrows, despised and rejected . . .* A different kind of Good Friday Experiment, a different order of ego-death, mystical experience, nature relatedness. I was moved to tears by the tree's suffering.

I had saved the best till last. Wachuma was my drug of choice; one part monk, one part punk, it seemed to fit my temperament nicely. There was something about the experience, its compression of the feeling of presence, that made it stand out from the other medicines. The Jesuit poet Gerard Manley Hopkins was fascinated by 'haecceity', a scholastic invention, literally meaning 'this-ness', coined to describe the irreducible determination of a thing, the *thing* that makes it *this particular* thing: 'the swanity of the swan, the monkitude of the monkey, the lotusness of the lotus' in Don Paterson's words.[5] The principle shaped the form and metre of Hopkins's poetry. Hearing it read aloud for the first time as a dull-eared, atheist teenager was like having my whole sensorium defibrillated.

> The world is charged with the grandeur of God.
> It will flame out, like shining from shook foil;
> It gathers to a greatness, like the ooze of oil
> Crushed.[6]

God apart, Hopkins became my poet. Decades later, and many years since I had last read him, somehow this nineteenth-century ecstatic rhapsody to presence informed my experience of *wachuma*: psychedelics teach us that everything inherits everything, excavating the buried strata of our internal geology, pointing out the infinite convergences of history, culture, species, as though we have struck on the set and setting of Being itself. I'm aware how grandiose and mystical that sounds, but that's in part the intoxicating effect of these intoxicated experiences. One might say I was 'on' Hopkins when I was on *wachuma*. Both summoned in me a similar quality of stillness somersaulting, of consciousness disclosing itself;

the exact same feeling-tone that had inspired me in my twenties to join a monastery.

But to make this feeling religious was, at least in my case, to mistake it; a case of trusting the teller over the tale. Because I got the same, or similar feelings from the anti-religious poetry of Wallace Stevens (who soon displaced Hopkins, as Bowie did the Beatles, in my teenage affections). I hadn't read Stevens for years, and then he showed up in the cactus garden. (I know: you wait decades for a poet and then two come along at once.) For Stevens, unlike Hopkins, Sunday mornings are better spent in the garden rather than in church, the perfect setting to conjure different 'ideas of order' out of the interplay of the natural world and the poet's mind. Like *wachuma*, Stevens's poems are really meditations on consciousness itself, and reading them its own roller-coaster, one moment uplifted by innuendos of the divine, the next plunged into bawdy bourgeois unbelief.

Which is all to say this was my own heady brand of psychedelic tourism, of colonial appropriation, of authenticity-questing. In the circumstances what else could I do? I understood so little of my 'setting' on this Andean hillside – the language, the names of the flora and fauna that grew around where I was planted, but also the dying indigenous traditions reflected in Alejo's pathos, or anything meaningful about the state of modern Colombian politics and culture, or narco-terrorism or gang violence – that my brain filled the blind spots with familiar fragments: scraps of English antiquity, some half-remembered lines of canonical Western poetry. It might have fallen on anything and yet, bizarrely, these random-seeming associations seemed to fit the occasion perfectly. Either that, or all this was just gobbledegook, like Michael Pollan, high and snagged on Oliver Wendell Holmes's 'a strong smell of turpentine prevails throughout', turbo-charged with a wisdom that won't survive the light of day.

It feels incumbent on us to give voice to these difficult-to-voice experiences. Asked to describe the experience of ineffability in psychedelics, the circuits of Apprentice Bard, Google's new language machine, go round in circles. *'Psychedelics are ineffable because their effects on consciousness, perception, and thought can be so profound and unique that they cannot be communicated through language. Instead they are ineffable.'* (It continues: *'Psychedelics should not be taken by anyone who has a history of mental health problems,'* somewhat contradicting the multi-billion-dollar medico-therapeutic project of the last few years.)

Apprentice Bard is just wrong about ineffability. Poetry is one way of expressing psychedelic-like experience. In fact, such experiences go to the heart of the best poetry, communicating our deep-felt need to have a language of our very own that captures the unique signature of the thing it describes, and the no less deeply felt need to share that with others. Really, all experience borders on ineffability – even the simplest falls into the cracks between conventions, or the space between one mind and another. Poetry's taproot is found in this gap. What we call the 'ineffable' in psychedelics is often a password for protecting something we think of as special, a way of resisting simple meanings, and resisting the functionalism that wants to 'make use' of everything.

I'm convinced that the kind of poetry that's pinned me to the spot over the years is its own kind of sorcery, in which the poem is both medicine and shaman (the philosopher John Stuart Mill found the poetry of Wordsworth the only effective medicine for his depression), the poet both receiver of things as they are and creator of them. There are infinite possible ratios between the self and the world, as various as every poem that was ever written or every psychedelic journey that was ever undertaken, from gaudy phantasmagorias of things as they never have been or could be, which spring like Athena from the Head of Zeus (to borrow Michael Hoffman's description of Stevens), to those that present the world as if our headsets appear to have been removed altogether, and we are left with a mesmerising fragility, the mereness of things as they really are. Here's the poet Don Paterson again on the shamanic art of poetic creation: 'To write a decent poem, I believe you need to have absorbed the truth about the fundamental ghost-hood of the human condition, our "double-realmed" twin citizenship status of being both now and eternal, alive and dead . . . You learn to flick a switch and strip away the name and utility from your shoes, or that chair, or that flower; it then returns to the set of the unfathomable, where its enfolded meaning allows it to blend and echo with other things.'[7]

And again, in lay neuroscientist mode: 'Poets' brains have a wiring error that makes them think words are real things. They keep trying to use words to magically influence reality, and to drag bits of it into the world that just won't go.'[8] All of which sounds like certain states in meditation or psychosis or psychedelic journeys. And among the arts it's not just poetry. You could easily make a case for calling the most inspired art of the last hundred years *psychedelic*. In different ways the beautiful manglings of Picasso, the otherworldly synaesthesias of Nabokov (read *Invitation to a Beheading*), the hyper-associations of Bob Dylan, the seemingly direct access to an infernal unconscious in David Lynch, or the way time opens up like a series of Russian dolls in the very different novels of Virginia Woolf and Annie Ernaux – all qualify, all created by artists who never went near a trip in their lives (Dylan an infrequent exception).

Which all begged the question: did I really need to do any of these trips? Well, did I?

'Why do you need to get high, Andrew?'

My father's voice. Though I had lived with him after my parents' divorce, I had not been particularly close to my father. That didn't simplify the impact his death had on me. I mentioned at the start of this book how he appeared to me the first night I took ayahuasca with Aurora, how I came to terms with something important in my relationship to him, how that incredible experience inspired the curiosity that led to this two-month road trip, which in turn led to this book. It turns out I was wrong. A year and a half later I hadn't come to terms with his death at all; rather I'd just begun to understand what those terms were. The deep feeling of lasting peace and connection in the cliff-top yurt in Big Sur gradually gave way to deeper, more enduring feelings of loss, agitation, fear – the exact same things I used to run from by getting high. It felt as though I'd never left home in the first place.

My father wanted me to be a doctor. After a decade of disappointing or worrying him in various ways I more or less fell in line (clinical neuropsychologist was just about 'acceptable'). But Dad's death ended the contract, the unspoken agreement to give him what he wanted; to be a doctor, to *not get high*. At that level my road trip had little to do with investigative research, and everything to do with the freedom to reclaim things – good and bad – that I'd left behind, to sink in the mud. Psychedelics forced my head into the mud, and one thing I found there was that some of the anger I felt about conceding to his wishes got transferred onto the medical profession itself; that my prior was to see the shortcomings in medicine before its goodness: its failure to do anything to stop his sudden death, or slow down Danny's ALS, or help my daughter, or make much material difference to the lives of the hundreds of brain injuries and neurodegenerative conditions, the griefs and traumas and depressions I'd encountered as a clinician.

Of course, this personal bias in some way shaped my intellectual resistance to the medical model's claims about psychedelic treatment. So that for every headline about the 'miraculous effects on depression', every tweet about the 'wonders of synaptogenesis', a voice in my head automatically countered: the flaws in psychiatry's world view, the problems of inference in neuroimaging, or the dark incentives of corporadelia. The voice was reasonable – it just had a little extra spice in my case, an accent that was a little too quick to downplay the reality of progress and potential, because under it all, but also attached to it in the way a tumour is said to be 'attached' to a vital organ, was the deeper need to break with the authority of my father. Now that he was dead, the opposition lived on in me.

To put it simply, my big psychedelic breakthrough was that I had had a breakdown. Meaning I finally recognised what had gnawed at me for decades without my feeling it, giving in fully to it – *dying*, you might say – so that I might recover from it. Or, more New Agely, I'd given birth to a 'child', as the Kogi divination had foretold, a child because it was a reality that was always already inside of me. Yes, it's a cliché, but that's also precisely the point: we can't have the truth without the muck that's ingrained, the baby without the bath water.

There it was, what I'd been looking for all along: my breakthrough, my diagnosis, myself as a miniature case study in the first person. You may say this father complex was the single amalgamated insight, the ur-insight, *caused* by all the psychedelic journeys, and there's no counterfactual available to refute that now – I couldn't untake the drugs. Except the way all this bubbled up and cohered over the different trips, piecing itself together laboriously, accidentally, murkily, but somehow perfectly, while I watched so many dragons and rainbows and eels, kept hugging total strangers, kept making a run for the hills when there were often no hills to run to, no legs to do the running, was, I thought, more suggestive of the slow art of life. Unintoxicated life, that is.

The next day began at 3.30 a.m in the foothills of Gelarus, the top of which could be seen through the night wearing a ruff of fog. The ceremony would follow the tradition of the Q'ero Indians, whom Alejo had lived alongside in the Ecuadorean Andes. Beginning at dawn, the Q'ero would spend the day walking in the mountains for the duration of the drug effects, thirteen hours or so, stopping at intervals to commune with nature, sing songs, make prayers and invocations.

With Alejo as our sentinel, we walked ponderously in single file through thick brush, which gave way to a copse of black oak that lined the sides of a valley on either side of a small brook ('copse' and 'brook' – indomitable Englishness! I couldn't help but translate natural patterns into the priors of an English countryside). This same small brook fed the Putumayo that fed the Amazon, a thousand miles south-east.

We climbed down the steep valley bank and, as we descended, I felt the brook rise up to meet us. I literally felt it: a deep bodily knowledge, that the brook was always rising, making a gift of itself, whether or not there was a *wachuma*-steeped mind to notice it.

At the bottom of the bank Alejo told us that the black oak we were about to seat ourselves under had been there for over 400 years. Recent erosion of the soil caused by climate change meant there was hardly any ground left to keep it upright. These were its last days on earth.

We sat a while and listened to the brook. Alejo played his kalimba, a thumb piano mounted on a hollow-skinned base which allowed him to drum and distort the melody's resonance by opening and closing the base like a wah-wah pedal. The arpeggios made a descant over the music of the brook; the kalimba's bass made low, wah-wah-like harmonies with the wind. 'The charm of music,' writes Jankélévitch, 'is to make a poet of every listener.'⁹

Looking up at the sun streaming through the branches of the dying oak, I discovered I had been crying for some time, as though crying itself was a natural background state, life's drone or bass note. Only this crying was without a hint of grief. Crying with a freshly discovered freedom. Crying for the first time in my life at the unaccommodated elegance of something. There was sadness on all sides: no ground left for the oak to stand on, for any of us; everything was going to die or disappear. But in that moment there were no such thoughts – those would have required history, time, concepts, a self to leverage them, and they had all withdrawn. No, it was spontaneous crying, beauty's reflex, a body finding its tune. Crying is the purge of *wachuma*.

The kalimba fell silent. A blackbird took over, as though responding to its mate. There are certain species of birds whose young can no longer learn the songs of their lineage because they are so close to extinction. The birds' native woods have fallen silent. So ornithologists had taught the orphans their own song with digital recordings. These young birds now sing songs from the seam of technology and nature, from the join of life and death. Like the notes used to wake Lear from his grave, or the rooster heard by St Peter (San Pedro) that augured his fatal betrayal, or the silence of the canary that saves a miner's life. Singing. Crying. Singing.

> *Salta de un cerro a otro con tu Espíritu*
> *Lagarto Lagartito.*
> *Vuela con tu mente, con tu fuerza*

Tus poderes Wachumita.
Tú no eres del Cielo pero vienes desde arriba
Mándanos tu remedio, tu corazón de alegría.
　　　'Ikkaruna Cantos de Buen Vivir', Felipe Kilakeo

Jump from one fence to another with your spirit
Little Lizard.
Fly with your mind, with your strength
Your powers, little Wachumita.
You are not from heaven but you come from above
Send us your remedy, your heart of joy.

Alejo was singing a dark, complex, witchy song over a manic drum to a circle gathered round an open fire, in an otherwise pitch-black *maloca*. The song had its roots in the now extinct pre-Columbian Chavín culture. Chavín songs are intentionally difficult to decipher, music that in itself is, says Rebecca Stone Miller, 'intended to transport the viewer into alternative realities'.[10] Psychedelic, in other words. The music was as unsettling as the moment by the oak had been serene, and we kept on crying all the same.

Its havoc wrought, the song came to an end. The feeling changed key once more. The elder cousin took up a guitar and sang a *vallanetto* to his brother called '*Los caminos de la vida*'. It was a traditional lament which was turned by his tremulous baritone and our molten sets into the most starkly melodious love song imaginable. Both of them were crying, the war over. Then, kneeling in front of the fire, they held one another, running once murderous fingers through each other's hair. And, watching from the edge, Jorge, the autistic boy, on three full cups of medicine, neither blank nor tender, but quizzical, like an anthropologist trying to apprehend something new, this Amazon of tears rushing past him, of people remembering how to get wet.

In terms of the history of the West it's not so long since tens of thousands would gather under an open sky and weep their way through the latest tragedy of Euripides or Aeschylus, feeling themselves implicated and transformed by the poetry of suffering. It was art that disclosed life, made it manifest. Though there have been echoes of its public force in the centuries since, gradually a way of seeing has been obscured by generations of reason, first in the form of religion, then as science. The poets, musicians and contemplatives remained to resist the doctrinaire with the beautiful, and luminaries in my own field like Luria and Sacks still listened. But there was hardly anybody left sitting round the dying fire to cry with.

'Art has failed us, but now there's a new Strong Poet: DMT.' The words of that bastardising bard of bombardment himself, Terence Kemp McKenna.[11] I really do wonder what psychedelics can teach us that we – collectively, unconsciously – don't already know, in some way, from art. Know and have forgotten, know and can't otherwise recover. I wonder if the learning might not be the other way round; that the Psychedelic Renaissance, which has to date been owned by clinical science in the West, might pupil itself in art and aesthetics – alongside history, culture and the rest of the humanities – and treat the trip not as an experiment or a therapy but as a poem, a drama, a dream. It's the way that psychedelics lead us to knowing that's different, and the feeling of knowing they allow, because for those few hours each of us turns into poetry.

The uncommon literary form, the 'century', consisting of a hundred brief *sententiae*, distinct but related by a common thread, was found congenial to the meditative mode of thought of early Christian monks such as Evagrius Ponticus and Maximus Confessor. The following is an iteration of this ancient form, a stoned attempt to show what poetry might say, has already said, about the 'ineffable': a psychedelic century, if you like.

1. Psychedelics show us we must be artists in all things. (*After Lowell*)

2. Psychedelics teach us that 'he who dwells in darkness like a spirit shall own all'. (*After Hofmannsthal*)

3. Psychedelics show us that death is the mother of beauty. (*After Stevens*)

4. Psychedelics teach us to be self-reflecting . . . and never look in the mirror while you're on drugs. (*After Izza*)

5. A psychedelic journey is a few hours spent in the library of the universe, the opportunity to return late books or, for the more ambitious, to re-catalogue the whole. (*After Santi*)

6. Psychedelics re-enchant us.

7. Psychedelics also show us enchantment is an early developmental stage, that disenchantment is the adult in the room.

8. Psychedelics are a tsunami whose lesson is, 'Surf, or be annihilated!' (*After Oberon*)

9. Psychedelics break the skin on the pool of the self. (*After Heaney*)

10. Psychedelic revelations are not the revelations of beliefs, but the precious portents of our own powers, and the still more precious portents of our humiliations.

11. Psychedelics are bitter herbs that sweeten the world's flavour. (*After Oberon*)

12. Psychedelics show us that our development depends on us becoming connoisseurs of chaos. (*After Stevens*)

13. Psychedelics show us our lives are split between the sublime and the ridiculous. Real 'integration' is knowing both at the same time.

14. Psychedelics teach us to fear people who are not on drugs. (*After Palmer*)

15. While we might be governed by Enlightenment reason in our public institutions, at home we are insatiably Romantic. (*After Charles Taylor*) Psychedelics teach us that both are equally necessary . . . and dangerous.

16. Psychedelics show the drunkenness of things being various. (*After MacNeice*)

17. Psychedelics are beyond the reach of the smartest phones. And while high, it's smart to have your phone beyond reach. (*After Hakan*)

18. Psychedelics show us that, even though reality may be a controlled hallucination, we can still get at truths of who and what we are and what we might be. (*After Heaney*)

19. A psychedelic trip should not mean but be. (*After MacLeish*)

20. Psychedelics are the opposite of most drugs, in that they can make you feel much worse when you take them and much better when it's over. (*After Dmitri*)

21. Psychedelics are a behaviourist's nightmare: reward and punishment can't be operationally distinguished.

22. The nobility of a psychedelic journey is a violence from within that protects us from a violence without. (*After Stevens*)

23. Psychedelic journeys are like a sequence of unimaginable sexual positions you will never get to try out. (*After Hakan*)

24. Psychedelics allow us to remember the dead while they're still alive. (*After Flore*)

25. Psychedelics are nature's Rorschaches. (*After Santi*)

26. Psychedelics are fucking Windex for your outlook. (*After Frank and before Prince Harry*)

100. Psychedelics change the laws of time, geometry and . . . arithmetic.

The fire was all but done. The Colombians had gone back down the mountain to eat fried trout soup and plantain fritters made by Alejo's mother. As was the custom at the end of many of the ceremonies these past weeks, Alejo had closed it – a little too early for my chemistry – with a momentous speech, each of us pinned to our seats by words whose actual meaning galloped unharnessed out the *maloca* door. The one thing I did retain was that – as with Taita Flore – there was repetition of the urgent need for the Chacruna, Bridge People, and – looking unblinkingly in my direction – this book should be in service of that.

The three gringos didn't want our journey to end. Hakan was going to take *wachuma* to Istanbul; Palmer would introduce it to Tibetan Buddhism; I would share it in private with my neurology patients. Three middle-aged men lying around the fire's remaining embers, dedicating our lives like teenagers to the afterglow. The conversation petered out.

After their first night of ceremony together in Sibundoy, Hakan had told Palmer, whom he had only met the day before, that among his visions that night he had the recurring 'taste' of Palmer's suffering. It had brought about an accelerated intimacy between them. Now, just when it seemed most natural for us to *go-bro*, hug, congratulate ourselves on the completion of our trip-tour and head off for supper, Hakan returns to it, first of all circumspectly, like trying to get at a fishbone in his throat. 'Don't you think you can be a little precious about your suffering, Palmer – a bit of a pain hoarder?'

Even Hakan's gentler tones are shockingly direct by Anglo-American standards. Then, after Palmer says nothing in response, Hakan lets his vigour and bluntness take over. 'Why is it you spend your life getting fucked up on psychedelic drugs? What is so unbearable about your life? You act like you're on some great cosmic quest, but from where I'm standing it looks like you've spent your life looking in the wrong direction.'

They barely knew each other.

Palmer was nailed, and he knew it.

'I mean, what the fuck happened to you? Come on, this is it: your last chance to be honest before you turn to stone.' Interrogation, tough love, a different kind of psychedelic-assisted therapy, like he's turned into his own father trying to break a Kurd with his cudgel-like

301

questions. And as Palmer fails to find any words in response, it's towards his own father that Hakan's anger now turns: the source of his migration to London, his shame, his self-evident brutality. Palmer and I listen as Hakan sobs his heart out into the fire.

Which means it's my turn to take up the baton of self-disclosure. I tell them about my father, and the new story that had finally broken through an hour or so before. Hakan and I cry a duet, the lachrymal equivalent of 'collective effervescence', the ashes smouldering with our tears. Until Palmer can no longer hold himself together.

He spoke in a voice I'd never heard before: high-pitched, trembling; eyes averted.

It was both shocking and familiar, extraordinary and mundane. As soon as it was said, everything I knew – the paranoia, the delinquency, the privacy, the questing, the cynicism – was constellated in a different way: the story, properly told at last, open-heartedly around a dying fire, melting ancient defences, seeming to reorganise him . . . (It reminded me of watching new dendrites flicker into life on YouTube, back in London, before the whole journey began.) Its climax, he told us, had come a few months ago when, after thirty years of silence, he looked up his mother online with a mind to finally reach out to her. She was the principal of a large public school in the Midwest, her résumé posted on the school website. It listed her many, considerable achievements, ending with her biggest accomplishment of all: being a mum to 'two wonderful adult children', one a lawyer, the other a nurse. These were her stepchildren, Palmer told us through excruciated sobs. Her only real son – Palmer himself – had been written out of the story of her life.

'There! You did it! I swear I saw a blackbird fly out of your chest as you spoke!' exclaimed Hakan, putting an arm around him. 'The shame is hers, Palmer, not yours. You shed it with your tears.'

Wachuma: the most cinematic of all the psychedelics.

In that week-long retreat of disappearing traditions and reappearing histories, we made the commitment, as sentimental as it was sincere, as unlikely as it was unavoidable, to grow old together.

Epilogue

Psychedelics are perceptual DJs, default mode modulators, isness-mongers, mystical experience machines, moral enhancers, genius-hacks, capital unicorns, the second-third-fourth generation gambits of humanity 3.0. They are also raw organic purgatives, MRIs of the soul, inter-dimensional grandmas, thisness-singers, anti-colonialists, messages in a bottle washed up from an alien sea. They will make you happy, green, unafraid, liberal, intelligent, creative, enlightened, ready to meet death. They will also make you uncertain, terrified, crazy, Republican, anarchic, delinquent. Unless, as a few old heads maintain, that's all a crock of shit and they are none of the above, but liars, fabricators, menaces, makers of mischief, wonky wreakers of weird . . . (For all their ineffability psychedelic descriptors are seemingly endless.)

Anyway, you decide.

In the last few years in the West there has been a renewed attempt to make (or re-make) and monetise models out of the usage of psychedelics, and by models I mean ways of focusing, directing and deploying their effect for a particular end. The convention is to separate a medical from a spiritual model, but these are commonly subdivided: medicine into neuroscientific and psychological / psychotherapeutic, spiritual into New Age or indigenous / shamanic, so that even within one broad category there are radically different ways of seeing the world, or thinking of the 'self', or understanding the meaning of a psychedelic experience. When even the basic languages are so different it's no wonder 'conversation' is hard and factionalism abounds. My experiences suggest that, whatever the model, something always gets lost, neglected, simplified or confused. Often the models themselves get mashed up: churches talking neuroscience and income per capita, psychotherapists saying the medicine is the real therapist while applying for patents for their playlists, and as for science, one commentator recently described the experience of reading a psychedelic paper as like asking ChatGPT to write an article about psychedelic neuroscience in the style of Deepak Chopra.[1]

And in a way a mash-up is no bad thing. To date the Psychedelic Renaissance has largely been a product of the synergy between finance and science. But, as my experiences suggest, a real renaissance is a mash-up that must include the humanities, which is to say it includes different perspectives on the values and meanings of experiences, and takes as crucial the cultures from which they emerge. Science, technology and business are part of the effort to 'instrumentalise' psychedelics in one way or another. The humanities, by contrast, are, according to one of their leading theorists, Hans Ulrich Gumbrecht, a way in which we render complex experiences *more* complex, by opening up other perspectives, challenging existing interpretations with inconvenient facts, unlikely findings, counterintuitive insights. Correspondingly, as Nicolas Langlitz has recently suggested, psychedelics offer 'an opportunity for conducting experiments not just with the drugs themselves but also with the research practices through which we come to know their effects'.[2]

More broadly, you might say they offer the opportunity for a *real* renaissance in the way they are thought about, requiring us to include art, music, philosophy, history, culture – alongside the traditional approach of science (and the role of corporate interests) – if we are to understand the true nature and implications of the psychedelic experience in all its particularity. As Langlitz suggests, only this kind of integrated thinking will work, for we who live in liberal democracies require and lack the guard rails of long-established traditions. And with respect to those traditions, our thinking must include, as Bia Labate and the Chacruna Institute among others have persistently advocated, an ongoing dialogue and material exchange with experts from indigenous traditions – like Taita Flore and Alejo – from which the plant-based medicines originate. Meanwhile our own experts, our scientists that is, might foster greater curiosity about experiences themselves, alongside respect for the wondrous and enigmatic (however far-flung they may first appear) and, in a field that is, despite appearances, sometimes more like quantum mechanics than ortho-dox neuroscience, recognise the unavoidability of observer effects.

Which brings us back to hype. I admit, it's hard to resist. My jour-ney was *the trip of a lifetime.* In many senses I'm still on it, destined to be catching up with it in different ways for years to come. It's left me convinced that one way or another, the drugs will clearly play a sig-nificant role – or roles – in our culture's (and others') future as the twenty-first century unfolds (though it's also clear they are unlikely to feature as much in that future as stimulants such as coffee, nicotine and amphetamines, and depressants like alcohol and opioids, will). Returning to one of our original metaphors, the hero's journey, it turns out that our hero is more complex than most, at times more Hyde than Jekyll, the drugs symptoms of our need for the irrational as much as they are an instrument to treat it. For all their promise, psychedelics are not a straightforward antidote to the mental health epidemic, as leading researchers in the field like Rachel Yehuda have pointed out. Personally, I've never met a clinician who characterises mental disorders as a problem to be 'solved'. To my ear, antidotes and solutions sound more like the creations of venture capital, whose model of reality is premised on funding companies with impossible goals, and whose raison d'être is the creation of flywheels for endless growth. Rather it seems as though two extraordinarily complex phenomena – psychedelics and mental health – have been simplified and conflated in a public discussion in which individualism, progress and prosperity are the undertow for much of how we think about ourselves. In reality our mental health, like our physical health, is in the long term irreparable, our suffering unavoidable, and the extent to which their 'avoidableness' is overstated is directly associated with the way that our 'failure' to avoid them is pathologised.

This has always been the case, but perhaps it's *more* true now than ever before. What can the difference between mental 'health' and 'ill-ness' mean when it's normal for people to spend a huge portion of their day more or less at the mercy of algorithms that disincentivise our appetite for real experience, hook us to a screen, seduce us to spend our time and money on one unsatisfiable fantasy or another? As far back as the 1950s William Burroughs prophesied that the only escape from technology's control is by considering 'our entire gadgetry as junk . . . apomorphine'.[3] He was referring to his own particular drug of choice, but his meaning was far-reaching: technology is the new opium of the masses. By the 1990s Terence McKenna predicted that 'the drugs of the future will be computers and the computers of the future will be drugs'.[4] Well, here we are in 2023. Individuals may change or not, may de-canalise their minds with therapy, meditation, walks in nature and drugs; meanwhile our digital environment does the opposite: it's *all* canal, digging itself deeper with each click by algorithms designed to make our brains more predictable, less our own. In this landscape, hype is the air we breathe, the signature symptom of an exchange rate – between the real and the virtual, between truth and misinformation – gone haywire. Day by day the chances of something being what it says it is are low and getting lower, and in a culture that demands the narcotic of novelty more than anything else, this has to affect the way we talk and think about psychedelics. Even this rant is part of it, a kind of over-hyped anti-hype, which is why it's time to stuff a sock in it.

This was the end of a journey made up of 'journeys', a little over one every other day for two months. I had amassed no more than several hundred hours of tripping, still very much a beginner by conventional standards of expertise. Other than those documented, I had several experiences with the 'N,N' variant of DMT, mainly in the form of smoked *mimosa hostilis* (outrageous visual extravaganzas but, in my case, emotionally empty). I was led in a traditional Mazatec *hongos* cere-mony with the daughter of one of Marina Sabina's Twelve (a simple, beautiful ceremony in a church-like tin hut on a mountain from which I made repeated, unsuccessful attempts to escape), and joined another Mazatec ceremony with a tea made from *salvia divinorum* (subtle, unremarkable, but that night I had the most compellingly strange dreams of my life). I also experienced Anil Seth's Dream Machine (stroboscopic light beamed on closed eyelids), and a Mexican *temazcal* (a traditional sweat lodge), both without any pharmacological support and both enough to stimulate hallucinations stronger than some of the chemical ones. Heat and light in certain doses, enough to persuade our brains of the psychedelic quality of drug-free reality.

Personally, the net effect of the ten trips was uncertain. The subjective toxicology report read that I was in good health, there were no obvious physical side-effects other than a dependence on chewing gum, fizzy sweets and *rappe* (addictions 'curated' rather than cured by the medicine, I would argue). The first few weeks after I returned to Britain I was more blithe than usual. The happiness had a dissociated feel, like a car accelerating down a hill, its driver unsure whether the engine is running or has cut out altogether. And during that time my mental capacities had the feeling of tyres spinning in snow. But soon enough attention, planning and happiness reverted to their pre-trip baseline. I returned to work.

The existential effects, however, were of a different order. Here the experimental report points to a trade-off between on the one hand the most intensely enchanted nights of my life, of astonishing transformations, of new friends, of love abounding, and on the other the sense of extraordinary experiences piling up on top of one another to the point where they consumed one another's meaningfulness. *Chief Cannibal*. Whatever that meaning was it bore little relation to my 'intention', which from my limited experience was usually no more important than the way a teddy bear is important to a child heading into the night. But no less important either. I would say that collectively the trips broke my personal version of what Paterson calls 'the fear-of-fear-of-fear vortex'.[5] I might also say things like, 'It's the road rather than the destination that matters', 'It's the trip itself that's important rather than its meaning', that 'We are more verbs than nouns, a process more than an operation'. But I'd also want to ask myself why I had to edge mental annihilation to learn such commonplaces, when, for example, I could have more easily and less riskily read much of the last few sentences in Paterson's recent memoir.

It's now several months later. The mind I have is recognisably the same as the one I left with. I don't have the feeling of having been cut in two as I did that first time with Aurora. My personal beliefs, my spirituality, my personality are more or less the same inconsistent soup they were before, though there have been subtle, significant changes in its flavour. I'm more aware of that part of me that always needs to run. It's not cured or healed, just *seen*, and seeing marks the beginning of tolerance. Equally, other viewpoints are less threatening, my own convictions looser. I have new (infrequent but regular) urges and desires: to wear kaftans, to do my own plumbing, to work in hospice care. Friends say I'm a little less impulsive, a little more grateful. After years of threatening to write, there's finally been a demonstrable shift in my creative output, this book being one piece of material evidence (some of it was written while micro-dosing with LSD, other bits on mushrooms; maybe the discerning reader will be able to tell).

For all my carping about hype, I'd be happy to go on record with the main lessons I have learned. Psychedelics have taught me that every experience is far richer and more unlikely than I normally give it credit for. They have strengthened my faith – not in a divine order or a benevolent universe and its manifesting potential, but a natural faith that any situation can be managed. On a good day such simple, humble lessons have been enough to re-enchant the world, even though it's been many months since I last sat in ceremony. I'm in no rush to plan the next of the trips – psychedelics are the epitome of *too much information*. In any psychedelic future, I would limit my consumption to ceremonies with known and trusted facilitators: as a seasonal means of reflecting on where my life had taken me, perhaps – where I wanted it to go, or how best to tolerate the thwarting of that want – and as an occasional reminder of the extravagant cosmic strangeness of everything I am too preoccupied to see otherwise, including litter bins. Whether I take psychedelics or not, I want to keep at least a toe in both worlds: not so grounded that I feel my life isn't worth describing to anyone, not so far out that I forget how to relate to the things and people around me.

And I recognise that my experiences, however broad and various, are from a singular perspective. As much as I've tried to wrestle them into a more general applicability, they remain inescapably personal. My preoccupations, my history, my interests, my set – neuroscience, meditation, poetry, continental philosophy, the Beatles, Johns

Hopkins, Sam Harris, Annie Ernaux, the Don Paterson book I'm reading right now, my daughter's hair, my rickety hip, the boy blinking too much on the Underground yesterday, and all the rest – are reference points I fold into understanding my psychedelic experience. Which is itself a self-portrait – white, male, over-educated, Western, middle-class – but in a convex mirror (to borrow from the poet John Ashbery). It just so happens that some of my references overlap with current medical research; another person with a different history and a different set of preoccupations would have charted a completely different route through the same territory.

Which is to say that perhaps the whole point of a psychedelic experience is to taste your entire life in a way you didn't know was possible – the unbelievable, terrifying, thrilling story of how you were born, the parents you inherited; how the *you* that you are took shape across all those different settings – the formative early years, your first heartbreak, the first time you took drugs, the birth of your child, and on through life, till the hour of your death – exposing the line that threads through all the decisions and accidents, the tangle of prejudice, ignorance, regret, bliss that somehow stitches you together. That's the therapy: psychedelics show you that you are the line of string threaded through the Minotaur's maze, you are the trail of breadcrumbs in the wood. In other words, you yourself are the way of finding home.

At the beginning of the book I suggested the metaphor of the multifaceted gem for the complex perspectives of psychedelics. But I've outgrown this metaphor, or better to say, psychedelics have slipped its net. A jewel is man-made, but *the* most fundamental lesson of my various trips is that these drugs defy our attempts to design them. Also, implicit in the jewel metaphor is the idea of a single definite form (albeit with myriad facets), and I'm not convinced that suits psychedelics.

Water, though, has no such form. While we might understand its chemistry through and through, or the behaviour of its molecules at different temperatures – the science of water, in other words – at the level of experience it's a substance that takes the shape of whatever contains it, while being impossible to observe without some container.

I've come to think of our psychedelic journeys in a similar way, as being like water. So whether we've got a mental health diagnosis or believe we have trans-human potential, whether we're tripping in a therapeutic 'living room', a Caribbean retreat or a *maloca* on an Andean mountain-top, whether we have embarked with a very specific intention, or no intention at all, whether we are accompanied by a Bach Mass, a Shipibo *icaro* or 'Strawberry Fields Forever', the water – the journey, that is – just keeps on changing form, bringing our lives to life in unfathomable ways, including all that we are, which is far more than we know.

It's not too dramatic to say that in the current moment we find ourselves on the edge of a psychedelic cliff. The shape of the water below looks like us at our worst: hyped, deludedly self-important, short-sighted, greedy, prejudiced, abusive, enthralled by technology that's beyond our control. (It's not too cynical to suggest that greed and self-importance, as much as progressive liberalism, are driving legalisation, and legalised psychedelics will change the shape of our container again, in ways we can't predict.) We are in danger of taking something we barely understand, something that holds huge promise for changing our perspectives – on mental health, social justice, eco-logical devastation and general human flourishing – and turning it into a prohibitively expensive, medico-spiritual Disneyland. I'm sorry, Aurora. I guess you won't like the sound of that, but on that note my various experiences only confirmed my initial scepticism: to para-phrase Einstein, we can't solve the problems we've created from the same level of consciousness from which we've created them. We have to change before psychedelics can really change us.

Five years on from the publication of *How to Change your Mind*, Michael Pollan's request for thoughtfulness and creativity in construct-ing our ceremonies and therapies is more apposite than ever. But in a more fundamental way, we lack a whole mythology around the safe and beneficial use of these plants and chemicals. As it is, we create our contexts on the fly, give them the jargon of science or religion, when most of the churches and clinical research programmes are barely a decade old. Real myths – call them wisdom if you like – are created over centuries, and abound with meanings warning of the dangers of excess, of greed, of desire, of hype. Indigenous cultures are filled with such stories, and in my experience their stories originate in meeting the plants on their terms, rather than ours. So before we design our ceremonies, we must create stories of our own, the right stories to help guide us on the path: a hero's journey, told properly.

We – my daughter and I – set off a little after 6 a.m. while it was still dark outside. There was one spot in the park I had in mind, to the south of Richmond Gate, where you walk up a gradual incline through thick woods until forty miles of city – the whole of London from left to right like a sudden wipe on the screen – whoosh into view. That was what I wanted her to see.

She was better, but still fragile. There had been times during the last two years when things had been as bad as we could imagine, and then they got worse. Imperial had advertised a new psilocybin trial for participants with her diagnosis, but she was much too young, barely a teenager. Out of sheer desperation, I had even fleetingly considered a mad dash to a ceremony in Colombia. That was more than a year ago. She was still not out of the woods, but better now for sure.

We made our way by phone-light on a trail that led us through the silhouettes of sycamores and oaks. Nigel, my therapist, was convinced that the real spring for my interest in psychedelics was a desire to help my daughter when no help was to be found. Had she been a little older, a little stronger, I might have suggested a low dose of mushrooms. Instead, I hoped nature on its own would be intoxicating enough. As we walked I wondered aloud how even though everybody knew we couldn't survive without plants, most of us knew nothing about them. Even your average farmer knew next to nothing about the micro-organismic cycles of the soil in which he makes his livelihood. One article I'd read recently suggested that your average man—

'Like you, for example.'

'Like me, for example. He knows less about the trees and plants around him, what their uses are – whether this one's medicinal . . .' I shone the torch at shelves of mushrooms growing like gills on the base of tree.

'Fatal, more like.'

'Or if this one's edible' – pointing it in the direction of some blackberries, having reached the limit of my botanical knowledge.

'Duh.'

'He knows less than a five-year-old child from an indigenous Amazonian tribe.'

People alive only a few hundred years ago here in Britain would have known which shrubs would make a good rope, which decayed to make good rooting compounds, which tree was best to build the frame of a house, which dried shrubs could insulate it. That deeper knowledge helped them to thrive in all sorts of ways, practically but also spiritually.

The canopy above our heads threshed in the breeze.

'All I'm saying is, if we're to survive – and I believe we will – then we need a bigger sense of who we are and what we need.'

She gave me one of her funny looks. She hadn't heard me talk like this before.

Just before I had left Colombia I had received a message from Santi inviting me to join a ceremony Taita Flore was leading in Sierra Nevada. Two of the Kogi were coming, their first time drinking *yage*.

Both of them had sat on the ground outside the *maloca* the whole night, not purging, not dancing or singing, eyes trained on the mountains, where their brothers and sisters were likely to be shuttling between sacred sites, making their payments. The next morning Santi told me that neither of the Kogi reckoned much to Taita Flore's tea. 'For them it was like a four-hour headache. It got in the way of seeing the Mother.'

The Mother: the understanding that the world is already psychedelic, that the whole of life should be understood with the reverence and mindfulness of a ceremony. That way of seeing would probably always remain beyond the grasp of Little Brother.

Back in London the trail had petered out, but it was light enough now to see without a torch. The two of us waded through the last thick swathes of wet grass. I stopped a few feet ahead, and turned back as the sun came up to watch the city breaking over her face.

Acknowledgements

Thanks to my editors, Will Hammond at The Bodley Head and Karen Rinaldi at Harper Wave, for their corralling clear-sightedness, and their unfailing, shaman-like belief that I was the right man for the job. (On that note thanks particularly to Will for just listening when I told him that 'Hyperdelic Smash Tour' was the only possible title for the book.) Thanks also to the ongoing kindness and wisdom of Will Francis, my agent in London, and to PJ Mark and Ian Bonaparte in the US.

The book involved meeting a lot of amazing and amazingly strange people along the way; I am deeply grateful to them all – as with the drugs/medicine/sacrament, my life will never be the same having met you. Particular thanks to Aurora for showing me the way; to Oberon and his family of assorted fairies for lending me a boutique, bug-infested cabin on their property to write, and much, much more; to my dear friends James Lever and Molly McGrann, who tried to ensure the early part of the book retained a little cool; and most of all to Hakan and Palmer, my travelling companions slash trip sitters slash integration therapists slash experimental controls for large parts of my field research.

Notes

Introduction

1 Michael Pollan, *How to Change Your Mind: The New Science of Psyche-delics* (Penguin, 2018)

2 Mike Jay, *The Psychonauts* (Yale University Press, 2023)

Chapter 1: Psychedelically Naïve

1 Joe's Twitter handle is @DrJoeDispenza

2 Christiana Westlin, Jordan E. Theriault, Yuta Katsumi *et al.*, 'Improving the study of brain-behavior relationships by revisiting basic assumptions', *Trends in Cognitive Sciences*, 27:3 (2023), 207–332

3 Though, as Hardman pointed out, a commercial effort to develop the R-enantiomer of ketamine and do away with the levels of dissociation and sedation seen in conventional ketamine failed to meet its primary endpoint in a Phase 2a trial.

4 'Five Questions for Manoj Doss', *The MicroDose*, 22 November 2021

5 Ben Sessa, 'Interview for Spanish *High Times*. At Tyringham Hall, Bucks, UK, on December 8th 2014, whilst attending a MAPS-sponsored MDMA Therapy Training Course', https://soundcloud.com/bensessa/ben-sessa-9-12-14-uk-maps

6 Christopher Timmermann, Hannes Kettner, Chris Letheby *et al.*, 'Psychedelics alter metaphysical beliefs', *Scientific Reports*, 11 (2021)

7 Annie Ernaux, *The Years*, tr. Alison L. Strayer (Fitzcarraldo Editions, 2018)

Chapter 2: Mystical Experiences

1 Wilfred R. Bion, quoted in Thomas Ogden, *Reclaiming Unlived Life: Experiences in Psychoanalysis* (Routledge, 2016)

2 Tehseen Noorani, 'Containment matters: set and setting in contemporary psychedelic psychiatry', *Philosophy, Psychiatry & Psychology*, 28:3 (2021), 201–16

3 *Ibid.*

4 R. Carhart-Harris, S. Chandaria, E. D. Erritzoe *et al.*, 'Canalization and plasticity in psychopathology', *Neuropharmacology*, 226 (2023)

5 Philippe Ariès, *Western Attitudes Toward Death: From the Middle Ages to the Present*, tr. Patricia M. Ranum (Johns Hopkins University Press, 1974)

6 Bruno Barnhart, *Second Simplicity: The Inner Shape of Christianity* (Paulist Press, 1999)

7 R. Carhart-Harris and K. Friston, 'REBUS and the Anarchic Brain', *Pharmacological Reviews*, 71:3 (2019), 316–44

8 Dacher Keltner and Jonathan Haidt, 'Approaching awe, a moral, spiritual, and aesthetic emotion', *Cognition and Emotion*, 17 (2003), 297–314

9 Timmermann *et al.*, 'Psychedelics alter metaphysical beliefs'

10 Chris Letheby, *Philosophy of Psychedelics* (Oxford University Press, 2021)

Chapter 3: Bad Trips

1 Alexander Shulgin and Ann Shulgin, *Pihkal: A Chemical Love Story* (Transform Press, 1992)

2 Anne K. Schlag, Jacob Aday, Iram Salam, *et al.*, 'Adverse effects of psychedelics: From anecdotes and misinformation to systematic science', *Journal of Psychopharmacology*, 36:3 (2022), 258–72

3 Rachael Peterson, 'A Theological Reckoning with "Bad Trips"', *Harvard Divinity Bulletin* (Autumn/Winter 2022)

4 *Ibid.*

5 'Dr. Adam Gazzaley, UCSF – Brain Optimization and the Future of Psychedelic Medicine', *The Tim Ferriss Show*, 30 March 2021, https://tim.blog/2021/03/30/adam-gazzaley-2/

6 'Psychedelic Pedagogy', *Expanding Mind*, 13 July 2017, https://expanding-mind.podbean.com/e/expanding-mind-psychedelic-pedagogy-071317/

7 Will Self, 'A Posthumous Shock: How Everything Became Trauma', *Harper's Magazine*, December 2021, https://harpers.org/archive/2021/12/a-posthumous-shock-trauma-studies-modernity-how-everything-became-trauma/

8 John Gray, *Straw Dogs* (Granta, 2003)

9 Albert Garcia-Romeu, Roland R. Griffiths and Matthew W. Johnson, 'Psilocybin-occasioned mystical experiences in the treatment of tobacco addiction', *Current Drug Abuse Reviews*, 7:3 (2014), 157–64

10 Peterson, 'A Theological Reckoning'

11 'Cover Story', *Power Trip*, 1 March 2022, https://podcasts.apple.com/gb/podcast/cover-story/id1594675355?i=1000552567394

12 Shulgin, *Pihkal*; *Tihkal: The Continuation* (Transform Press, 1997)

13 *Ibid.*

14 *Ibid.*

15 *Ibid.*

16 *Ibid.*

17 'Psychedelic Science', *Expanding Mind*, 29 April 2017, https://techgnosis.com/psychedelic-science/

Chapter 4: The Substitute Trip

1 I. Hartogsohn, 'Cyberdelics in context: On the prospects and challenges of mind-manifesting technologies', *Frontiers in Psychology*, 13 (2022)

2 'How Much Does the Future Matter?: A Conversation with William MacAskill', *Making Sense with Sam Harris*, 14 August 2022, https://www.samharris.org/podcasts/making-sense-episodes/292-how-much-does-the-future-matter

3 Don Paterson, *Toy Fights: A Boyhood* (Faber & Faber, 2023)

4 Daniel Pinchbeck, *Breaking Open the Head* (Broadway Books, 2003)
5 Ed Prideaux, 'The psychedelic utopia is a lie', *UnHerd*, 7 July 2022, https://unherd.com/2022/07/the-psychedelic-utopia-is-a-lie/

Chapter 5: On Harmony

1 M. Kaelen, L. Roseman, J. Kahan *et al.*, 'LSD modulates music-induced imagery via changes in parahippocampal connectivity', *European Neuropsychopharmacology*, 26:7 (2016), 1099–1109
2 Iain McGilchrist, *The Master and His Emissary: The Divided Brain and the Making of the Western World* (Yale University Press, 2010)
3 Hazrat Inayat Khan, *The Mysticism of Sound and Music* (Shambhala Publications, 1996)
4 Daniel Stern, *The Interpersonal World of the Infant* (Routledge, 1985)
5 Shulgin, *Pihkal*
6 Inayat Khan, *Mysticism of Sound and Music*
7 Dr Oliver Sacks, 'Why the Brain Loves Music', Columbia University Lecture, 18 October 2007, https://www.youtube.com/watch?v=KWTRQhF70k0&t=4960s
8 *Ibid.*
9 Inayat Khan, *Mysticism of Sound and Music*

Chapter 6: Psychedelics and Meditation

1 This description comes from my friend and long-term meditator Michael Highburger
2 Daniel Goleman and Richard Davidson, *The Science of Meditation: How to Change Your Brain, Mind and Body* (Penguin Life, 2017)
3 Letheby, *Philosophy of Psychedelics*
4 Carhart-Harris and Friston, 'REBUS and the Anarchic Brain'
5 F. Streeter Barrett, M. W. Johnson, R. R. Griffiths, 'Psilocybin in long-term meditators: Effects on default mode network functional connectivity and retrospective ratings of qualitative experience', *Drug and Alcohol Dependence*, 171 (2017)

6 L. Smigielski, M. Kometer, M. Scheidegger *et al.*, 'Characterization and prediction of acute and sustained response to psychedelic psilocybin in a mindfulness group retreat', *Scientific Reports*, 9:14914 (2019)

7 'The Paradox of Psychedelics', *Making Sense with Sam Harris*, 28 June 2022, https://www.samharris.org/podcasts/making-sense-episodes/286-the-paradox-of-psychedelics

8 Pollan, *How to Change Your Mind*

9 *The Joe Rogan Experience*, 18 January 2019, https://www.youtube.com/watch?v=7MNv4_rTkfU

10 Letheby, *Philosophy of Psychedelics*

11 Henry Shukman, *One Blade of Grass* (Yellow Kite, 2021)

12 *Ibid.*

13 *Ibid.*

14 'Neuroscience Meets Psychology – Dr. Andrew Huberman', *The Jordan B. Peterson Podcast*, https://podcasts.apple.com/us/podcast/the-jordan-b-peterson-podcast/id1184022695?i=1000582548636

15 McGilchrist, *Master and His Emissary*

16 Willeke Rietdijk, 'A micro-phenomenological study of processes and mechanisms of insight meditation', International Symposium for Contemplative Research, Phoenix, AZ, 8 November 2018

Chapter 7: The Psychonaut

1 Erik Davis, *High Weirdness: Drugs, Esoterica and Visionary Experience in the Seventies* (The MIT Press, 2019)

2 Aldous Huxley, *Selected Letters*, ed. James Sexton (Ivan R. Dee, 2007)

3 *Terence McKenna's True Hallucinations*, directed by Peter Bergmann, 2016, https://www.youtube.com/watch?v=8MG5gFtZ3U8

4 Richard P. Bentall, *Madness Explained: Psychosis and Human Nature* (Penguin, 2004)

5 Donald Winnicott, quoted in Michael Eigen, *The Psychoanalytic Mystic* (Free Association Books, 1998)

6 Carhart-Harris and Friston, 'REBUS and the Anarchic Brain'

7 *Ibid.*

Chapter 8: 'You Will Have Another Child'

1 C. Arbeláez Albornoz, 'El lenguaje de las burbujas: apuntes sobre la cultura médica tradicional entre los kogui de la Sierra Nevada de Santa Marta' in A. Colajanni (ed.), *El pueblo de la montaña sagrada: Tradición y cambio* (Comisión de Asuntos Indígenas, 1997), pp. 181–204

2 K. Williams, O. S. G. Romero, M. Braunstein *et al.*, 'Indigenous Philosophies and the "Psychedelic Renaissance"', *Anthropology of Consciousness*, 33:2 (2022), 506–27

3 Gerardo Reichel-Dolmatoff, 'The Great Mother and the Kogi Universe: A Concise Overview', *Journal of Latin American Lore*, 13:1 (1987), 73–113

4 Pascal Boyer, 'Why Divination?: Evolved Psychology and Strategic Interaction in the Production of Truth', *Current Anthropology*, 61:1 (2020)

5 McGilchrist, *Master and His Emissary*

Chapter 9: The Evil Wind

1 Manuel Alvis, 'The Indians of Andaqui, New Granada [Notes of a Traveller; published by José Maria Vergara y Vergara and Evaristo Delgado, Popayan, 1855]', *Journal of the American Ethnological Society*, 1 (1860–1), 53-72

2 David Dupuis, 'Can Psychedelics Really Change the World?', MIND Foundation Blog, 6 August 2021, https://mind-foundation. org/psychedelic-technologies/

3 Michael Taussig, *Shamanism, Colonialism and the Wild Man: A Study in Terror and Healing* (University of Chicago Press, 1987)

4 *Ibid.*

5 *Ibid.*

6 *Ibid.*

7 Luis Eduardo Luna, 'Indigenous and mestizo use of ayahuasca: An overview' in Rafael G. Dos Santos (ed.), *The Ethnopharmacology of Ayahuasca* (Transworld Research Network, 2011), pp. 1–21

8 Taussig, *Shamanism, Colonialism and the Wild Man*

9 R. Dunbar, 'Religion, the social brain and the mystical stance', *Archive for the Psychology of Religion*, 42:1 (2020)

10 M. J. Winkelman, 'The evolved psychology of psychedelic set and setting: Inferences regarding the roles of shamanism and entheogenic ecopsychology', *Frontiers in Pharmacology*, 12 (2021)

11 Taussig, *Shamanism, Colonialism and the Wild Man*

12 *Ibid.*

13 *Ibid.*

14 Timmermann *et al.*, 'Psychedelics alter metaphysical beliefs'

15 David Dupuis, 'The socialization of hallucinations: Cultural priors, social interactions, and contextual factors in the use of psychedelics', *Transcultural Psychiatry*, 59:5 (2021)

Chapter 10: A Psychedelic Century

1 Erik Davis on Twitter, @erik_davis, 3 February 2023

2 'The Trouble with AI: A conversation with Stuart Russell and Gary Marcus', *Making Sense with Sam Harris*, 7 March 2023, https://www.samharris.org/podcasts/making-sense-episodes/312-the-trouble-with-ai

3 Keith Thomas, *Man and the Natural World: Changing Attitudes in England 1500–1800* (Allen Lane, 1993)

4 Marianne Moore, 'Poetry' in *The Complete Poems of Marianne Moore* (Faber & Faber, 2003)

5 Paterson, *Toy Fights*

6 Gerard Manley Hopkins, 'God's Grandeur' in *Poems and Prose* (Penguin Classics, 2008)

7 Paterson, *Toy Fights*

8 *Ibid.*

9 Vladimir Jankélévitch, *Music and the Ineffable*, tr. Carolyn Abbate (Princeton University Press, 2003)

10 Rebecca Stone Miller, quoted in Gray, *Straw Dogs*

11 Terence McKenna, 'Opening the Doors of Creativity', Lecture 1990, https://www.youtube.com/watch?v=R7tEDQFi7Po

Epilogue

1 Philip Corlett on Twitter, @PhilCorlett1, 4 February 2023
2 Nicolas Langlitz, 'What good are psychedelic humanities?', *Frontiers in Psychology*, 14 (2023)
3 Marshall McLuhan, 'Notes on Burroughs', *The Nation*, 28 December 1964, pp. 517–19
4 Hartogsohn, 'Cyberdelics in context'
5 Paterson, *Toy Fights*

Index

AM indicates Andy Mitchell.

Index

Andy Mitchell is a neuropsychologist and therapist. He has specialised in treating patients with rare brain conditions, head injuries and epilepsy, and in the application of mindfulness for neurological patients. As a therapist he has worked with people with a range of mental health disorders. Before entering medicine, his first degree was in English Literature at Oxford University. He is originally from Leeds.